The Red Notebook of Charles Darwin

The Red Notebook of CHARLES DARWIN

Edited
with an Introduction and Notes
by Sandra Herbert

BRITISH MUSEUM (NATURAL HISTORY)

CORNELL UNIVERSITY PRESS
Ithaca and London

First published 1980 by Cornell University Press.

International Standard Book Number 0-8014-1226-9
Library of Congress Catalog Card Number 78-74215

Composed by William Clowes & Sons, Ltd, London
Jacket illustration courtesy of the San Diego Zoo

Printed in the United States of America

Librarians: Library of Congress cataloging information appears on the last page of the book.

To Sydney Smith

Contents

Symbols and Abbreviations

[ ]	Darwin's addition
〈 〉	Darwin's cancellation
[ ]	Editor's remark
[. . . ?]	Uncertain reading
|	End of notebook page
e	Wholly or partly excised page

CR	*The Structure and Distribution of Coral Reefs* (1842)
Diary	*Charles Darwin's Diary of the Voyage of H.M.S. Beagle* Edited from the MS by N. Barlow (1933)
GSA	*Geological Observations on South America* (1846)
JR	*Journal of Researches* (1839)
VI	*Geological Observations on the Volcanic Islands Visited during the Voyage of H.M.S. Beagle* (1844)

Introduction

The Red Notebook is one of a series of notebooks kept by Charles Darwin during and immediately following his service as naturalist to the 1831–1836 surveying voyage of H.M.S. *Beagle*. It forms part of the collection of Darwin manuscripts at Down House in Kent, Darwin's former home, and, since 1929, a museum in his honour. The notebook came to Down House by arrangement with the Darwin family following Sir George Buckston Browne's purchase of the house for use as a museum.[1] It is a well-made but otherwise ordinary pocket notebook, measuring $6\frac{7}{16}'' \times 3\frac{15}{16}''$ (164 mm × 99 mm), leather bound with a metal latch, which still works, and, as the name suggests, red in colour, although the original brilliance has faded. The leather cover is embossed with a border design on both sides. The front cover of the notebook bears the initials 'R.N.', written on a rectangular piece of white paper. On the back cover is pasted a similar piece of paper with the identical initials and the additional phrase 'Range of Sharks', referring to an entry within the notebook. There is also an ominous epigram written in larger letters across the back of the notebook: 'Nothing For any Purpose'. All of these inscriptions are written in brown ink in Darwin's handwriting.

The pages of the notebook are of good quality paper, chain lined, and cut from stock bearing a 'T. Warren 1830' watermark. There are ninety leaves in the notebook or, in Darwin's occasionally irregular numbering, one hundred and eighty-one pages. Counting fractionally excised pages as half pages, seventy-five pages, or just over 41 per cent of the total, do not now appear in the bound notebook. All but twelve of these excised pages have been identified from among the general holdings of Darwin papers at Cambridge University Library.[2] In addition to the large number of excised pages, two other features of the notebook are worthy of notice with respect to its physical appearance. First, most entries are scored through with a single vertical line. Such scorings are common among Darwin manuscripts and were an indication to himself that he had made use of the material. Second, the notebook as a whole divides neatly into two parts on the basis of the medium in which the majority of entries were written and with respect to the direction in which they were written on the page. Up to page 113, entries are in pencil, except for a few later additions, and are written along the vertical of the page, that is, running across its narrower dimension. In contrast, from page 113 on, the majority of entries are in brown ink rather than pencil and are written across the horizontal of the page.

The obvious difference in appearance between the two parts of the notebook has a commonplace explanation. In appearance the first part of the Red Notebook resembles Darwin's field notebooks from the voyage, the second part the majority of notebooks kept after the voyage. Entries in notebooks from the voyage were typically made in pencil and written on the vertical of the page. In the field, pencil is obviously superior to pen and ink for writing. Also, for most right-handed people, a hand-held pocket-sized notebook is more easily written in when held vertically in the left hand, the palm of the left hand supporting the entire page, than when held horizontally, which leaves the right-hand edge of the page without support. Thus, so long as he was travelling,

Darwin wrote in pencil with the notebook held vertically. Once back in England, however, Darwin wrote while sitting at a desk and used pen and ink as well as pencil. At home he found it easier to write horizontally across the page.

This explanation for the difference in appearance between the two parts of the notebook is borne out if an attempt is made to date the notebook on the basis of content. Fortunately the first part of the notebook can be dated with some confidence on the basis of places named in the text. The first part of the notebook yields a perfect progression of place names corresponding to points visited by the *Beagle* from late May to the end of September 1836. The *Beagle* arrived at the Cape of Good Hope on 31 May 1836 and departed 18 June; page 15 of the Red Notebook contains the entry 'off Cape of Good Hope 70 fathoms 20 miles from the shore', and page 32 mentions the names of two prominent English residents of Cape Town, Sir John Herschel and Dr Andrew Smith. Page 38 refers to similar detail for St Helena, the *Beagle*'s port from 8–14 July. Later in the notebook, on page 45, reference is made to Ascension Island which the *Beagle* visited from 19–23 July, and on page 77 a reminder is given to 'Mem. Ascension'. On page 94 Darwin wrote "I now having seen Pernambuco" which he visited from 12–17 August. Continuing this series, page 99 mentions the Cape Verde Islands, a port for the *Beagle* from 31 August–5 September. Finally, page 107 refers to the Azores where the *Beagle* stopped for mail on 25 September, eight days before anchoring at Falmouth. On 2 October 1836 Darwin departed the *Beagle* at Falmouth for his home in Shrewsbury.

Exact dates for the second part of the Red Notebook are harder to determine. With no written itinerary available, one must look to different types of evidence. One useful guide for dating pages 113–181 are the names of individuals Darwin cited in his entries. Chiefly they are of scientific men active in London, most of whom Darwin had not met before returning to England. Thus on page 113, the first extant page in the second part of the notebook, Darwin referred to consultations with Richard Owen, then a young anatomist working at the Royal College of Surgeons, later the first Superintendent of the Natural History Museum in South Kensington. Since Darwin first met Owen in October 1836, the month Darwin returned to England, entries from page 113 on in the notebook are post-voyage in date.[3] Reference on subsequent pages to other London men—among them geologists Roderick Murchison and Charles Lyell, the geographer Sir Woodbine Parish, and the conchologist James de Carle Sowerby—substantiates this view. Further, the impact that these London men had on Darwin's work from the voyage provides evidence for dating certain passages. For example, at the end of January 1837 Richard Owen provided identifications of a number of Darwin's fossil mammal specimens from South America to Charles Lyell for inclusion in his presidential address of 17 February 1837 to the Geological Society of London.[4] Among the specimens which Owen identified was a llama-like animal, the *Macrauchenia patachonica*. This specimen figured prominently in the Red Notebook since the 'extinct Llama' referred to on page 129 was in fact the *Macrauchenia*. As the llama-like character of this specimen was unknown before Owen examined it, the entry on page 129 of the Red Notebook must have been written after the end of January 1837. How long after can best be determined by considering the context in which Darwin's remarks on the *Macrauchenia* were made. To do so it is necessary to consider Darwin's remarks on species in the second part of the Red Notebook as a unit.

The whole run of entries on species in the second part of the Red Notebook is important for the limited purpose of dating specific passages, such as those referring to the *Macrauchenia*. It is also essential for the larger purpose of establishing a date for Darwin's arrival at a belief in the mutability of species. On this last point it should be stated that while Darwin's observations during the *Beagle* voyage were fundamental to his work on evolution, his notes from the voyage do not reveal him to have been an evolutionist. He was at the stage of asking basic questions.[5] It should also be stated that there has previously been no fully satisfying evidence to document Darwin's own claim that he began to form his views on the species question in 'about' March 1837.[6] Undoubtedly the chief significance of the Red Notebook is that it provides such evidence. In the Red Notebook are found explicit indications that Darwin was ready to assert the possibility that "... one species does change into another..." (Red Notebook, p. 130). Equally important, Darwin's remarks on the species question in the second part of the notebook are sufficiently extended to allow one to characterize his position in some detail.

Darwin's remarks on species in the second part of the notebook are directed towards three general topics: the geographical distribution of species, a comparison between the distribution of species through time and through space, and the generation of individuals and species. The central theoretical notion to emerge with respect to geographical distribution is that of the 'representation of species' (p. 130), or what Darwin referred to in his autobiography as 'the manner in which closely allied animals replace one another in proceeding southwards over the [South American] Continent...'.[7] From this notion Darwin drew the tentative conclusion that such representative species as (to take his example) the two South American rheas had descended from a common parent (p. 153e). It is important to point out that in drawing this conclusion, Darwin chose to avoid a Lamarckian understanding of the bounding of species.[8] For Lamarck, species graded indistinguishably into one another. In contrast, Darwin perceived differences between even the most closely related species, a perception captured by his notion of representative species and confirmed by the judgements of taxonomic authorities.[9] The word he employed to describe this situation was 'inosculation', a medical term referring to the joining of one blood vessel to another. Because he saw species inosculating rather than grading into one another, Darwin believed at the time he wrote this entry that species change, or transmutation, must be produced 'at one blow' (p. 217), or 'per saltum' (p. 130). Transmutation and representative species aside, however, there are other statements in the notebook which bear on Darwin's more general understanding of the identity of species as correlated with geographical location and range. One such passage argues that 'new creations' of species are independent of the size of the land area inhabited by the species (p. 127). Other statements challenge by way of example the notion that climate entirely determines the distribution of species (pp. 128, 134e) or that species are perfectly adapted to a particular set of physical circumstances (pp. 129, 133). Behind such statements lie broad questions concerning the relation of the history of the earth to the history of life. Yet in these passages the tentative and empirical nature of Darwin's inquiries is paramount.

The second topic of interest with respect to the species question in the Red Notebook is the analogy Darwin drew between the distribution of species over space and over time. Darwin's statement reads: "The same kind of relation that common ostrich bears to (Petisse, & diff kinds of Fourmillier): extinct Guanaco to recent: in former case position, in latter time. (or changes consequent on lapse) being the relation.—"(p. 130) If we can simplify this statement by omitting the phrase 'diff kinds of Fourmillier' (on this see footnote 154 to the text), we have the following proportion: the common rhea stands to the lesser rhea as the 'extinct Guanaco' stands to the present-day guanaco. (The common rhea is *Rhea americana*, the lesser rhea the *Rhea darwinii* [*Pterocnemia pennata*]. [See note 149 to the text.] The extinct Guanaco is *Macrauchenia patachonica*, the animal identified by Richard Owen; the present-day guanaco is *Lama guanicoë*. [See note 152 to the text.]) The first ratio, between the rheas, was based on spatial succession, the geographical ranges of the two birds being contiguous. The second ratio, between the *Macrauchenia* and the guanaco, involved temporal succession, although the exact nature of the succession is not specified in the text. The common element binding the two ratios derives from the fact that both involved the replacement of one species by an allied species. Moreover, in context it is clear that replacement implied transmutation, for immediately upon asserting an analogy between spatial and temporal succession, Darwin referred to species changing. In doing so Darwin returned to a point we have noted earlier, namely his belief that allied species do not grade into each other. The *Macrauchenia* (or its cousin) must have 'inosculated' into the present-day guanaco, just as one rhea 'inosculated' into the other. For Darwin this ruled out the Lamarckian notion that one species gradually changed into another in response to 'degenerating circumstances' as might be caused, for example, by a gradual change in climate. However, even if Darwin did not embrace a Lamarckian mechanism for species change, he did share Lamarck's conclusion that present-day species were descended from earlier related forms.

A third topic taken up in the Red Notebook with general relevance for the species question was generation, or, in modern terms, reproduction. In the notebook Darwin dealt briefly with the particular issue of how he might regard individuation as occurring in the zoophytes, a now-abandoned grouping of plant-like animals whose most familiar representatives are the corals. (Red Notebook, p. 130) The technical nature of zoophyte generation was not Darwin's primary concern. He was chiefly concerned to see where zoophyte generation might fit in the general analogy he was drawing between the generation of species and the generation of individuals. The chief advantage of the analogy was that it gave Darwin an alternative to environmental explanations for the origin and extinction of species. Thus, on p. 129: "Should urge that extinct Llama owed its death not to change of circumstances; reversed argument. knowing it to be a desert.—Tempted to believe animals created for a definite time:—not extinguished by change of circumstances: . . . " Approached in this manner, "There is no more wonder in extinction of species than of individual.—" (p. 133) Although the claim is not made explicitly in this notebook, Darwin presumed that the complementary relationship might also hold, that the birth of new species might be understood by analogy to the birth of individuals.

Such are the major questions concerning species which Darwin addressed in the Red Notebook: geographical distribution, the relation between the spatial and temporal distribution of species, and generation. But can the passages in which these topics are discussed be dated to 'about' March 1837? Three kinds of evidence suggest they can. First there is the direct evidence from the notebook itself. As has already been mentioned, the second part of the notebook is post-voyage in date and references to the 'extinct Llama' and 'extinct Guanaco' on pages 129–130 derive from Richard Owen's work of January 1837. The important passages on species on pages 129–130, and very probably the whole run of remarks on species from pages 127–133, therefore date from January 1837 at the earliest. Equally important, other datable entries in the notebook are consistent with a March 1837 dating for pages 127–133. These entries include: (i) a reference on page 143e of the notebook to the 29 April 1837 issue of the weekly journal the *Athenæum*, and (ii) a reference on page 178 to the subject matter of conversations Darwin was having with the botanist Robert Brown in April and early May 1837.[10] (Since the entry on page 178 is the last datable one in the notebook, it also provides an approximate closing date for the notebook of May 1837.) The second source of support for a March 1837 dating of the main run of entries on species is the positive, although incomplete, correlation between the contents of these entries and Darwin's description of his original insight into the species question where he wrote that he "Had been greatly struck from about month of previous March on character of S. American fossils—& species on Galapagos Archipelago. These facts origin (especially latter) of all my views."[11] By the phrase 'character of S. American fossils' Darwin undoubtedly had in mind the similarity between past and present South American forms.[12] The relationship between the *Macrauchenia* and the guanaco which figures so largely in the Red Notebook was of this sort. Thus, on the point pertaining to South American fossils, the correspondence between the Red Notebook and Darwin's 'Journal' entries is exact. With respect to Galápagos species the correspondence is less revealing, for Galápagos species are not mentioned directly in the notebook. It may be that Darwin had in mind the Galápagos mockingbirds when he referred to the 'Calandria' or South American mockingbirds.[13] However, this is uncertain. On the evidence therefore, the correlation between the Red Notebook and Darwin's description of his insights of March 1837 is positive although incomplete.

The third line of evidence joining the Red Notebook to Darwin's insights of March 1837 derives from a comparison between the passages on species in the notebook with those in the opening pages of a notebook, labelled 'B', begun in July 1837.[14] In general, Notebook B carries forward discussions on species begun in the Red Notebook. Identical elements appear in Notebook B as in the Red Notebook but they are handled with greater assurance and, in certain cases, with the addition of new material. Compare, for example, the opening sentence on page 130 of the Red Notebook with the statement on pages 16–17 of Notebook B: "I look at two Ostriches as strong argument of possibility of such change; as we see them in space, so might they in time.—" Similarly, the notion of representative species reappears in Notebook B, although now in conjunction with the idea of isolation as a mechanism for species change. A series of entries on pages 7–10 of Notebook B reads as follows:

> Let a pair be introduced and increase slowly, from many enemies, so as often to intermarry—who will dare say what result.
>
> According to this view animals on separate islands, ought to become different if kept long enough apart, with slightly differ[ent] circumstances.—Now Galapagos tortoises, mocking birds, Falkland fox, Chiloe fox.—English and Irish Hare.—
>
> As we thus believe species vary, in changing climate we ought to find representative species; this we do in South America closely approaching.—But as they inosculate, we must suppose the change is effected at once, something like a variety produced—every grade in that case [it] seems is not produced?—
>
> * * *
>
> If species (1) may be derived from form (2) etc.,—then (remembering Lyell's arguments of transportal) island near continents might have some species same as nearest land, which were late arrivals, others old ones (of which none of same kind had in interval arrived) might have grown altered. Hence the type would be of the continent, though species all different.—

The greater sophistication of this treatment of the notion of representative species—its ennumeration of examples, discussion of the transportal of species, and indication of isolation as a mechanism for change—suggests that it postdates the passages on species in the Red Notebook. It is therefore reasonable to conclude that Darwin made his entries on the species question in the Red Notebook before he opened Notebook B—that is, some time before July 1837.[15]

Evidence for dating the important run of entries on pages 127–133 of the Red Notebook can now be summarized. As already mentioned, the dependence of some entries in that series on Richard Owen's work of January 1837 makes that the earliest possible date for the series taken as a whole, while the existence of Notebook B suggests July 1837 as the latest possible date for the entries. Thus, the entries must have been written during the six months from late January to early July 1837. At this point it becomes relevant to weigh the author's own word carefully, for in the absence of any evidence to the contrary it provides the best reason to assign these passages to one month rather than another in this period. Overall, in Darwin's characterization of this period, the month of March stands out, for it was then that he claimed to have come to his new view of species. Since the passages in question from the Red Notebook are clearly transmutationist, it is plausible to assign them that date. Contributing evidence from the notebook—the rough correspondence of the insights Darwin described having in March 1837 with the passages from the notebook, and the compatibility of two dates further on in the notebook (the reference on page 143e to the 29 April 1837 issue of the *Athenæum* and the reference on page 178 to activities of April and early May 1837) with a March dating for pages 127–133—supports this conclusion. The run of entries on the species question in the Red Notebook should therefore be assigned, however loosely, to March 1837. To this dating two qualifications must be stressed: that Darwin himself was tentative in dating the origin of his new views (to 'about' March) and that it is more than likely that various entries were written at scattered times. Yet, on balance, the evidence supports an approximate dating of March 1837 for the entries on pages 127–133 of the Red Notebook.

The obvious next question is: why March? Fortunately, if one takes the passages on species in the Red Notebook to coincide with the origin of Darwin's new views that question can be answered, for insofar as passages on transmutation in the notebook depend on new information unavailable to Darwin before March 1837, they depend on the identification of specimens from his collection by London zoologists.[16] These identifications were of two kinds: (i) identifications of specimens of unknown character and (ii) the marking off and naming of good species. The utility of expert opinion is obvious in the first case, as, for example, where Richard Owen's knowledge of comparative anatomy enabled him to identify fossil specimens where Darwin could not. The utility of expert opinion in the second case was less obvious, for the naming and marking off of species entailed considerable judgement on the part of the taxonomist. Yet while the process of species definition was in part arbitrary it was not anarchic. What ordered the process was the existence of recognized arbiters, or, in a word, specialists. The concept 'species' was in fact defined by such men as John Gould and Richard Owen as they went about their daily work. It was their office, not Darwin's, to name his specimens. Hence what one finds in the Red Notebook (despite its retention of common names for species) are traces of Darwin's reflections on his own initial observations in the field as these observations were ratified or extended by professional judgements on various specimens. One can see the effect of professional judgement in the insights he recalled as central to his new views. His first insight pertained to the temporal succession of similar species in the same locality. In Darwin's autobiography he cited the glyptodon-armadillo relationship to this end;[17] in the Red Notebook it is the *Macrauchenia*-llama replacement which figures.

Although the final word on the fossil mammals came from Richard Owen in both cases, Darwin's dependence on professional judgement is more obvious in the case of the *Macrauchenia*, for, unlike the glyptodon-armadillo relationship, Darwin had not formed any judgement of his own on the specimen before learning Owen's opinion.

Professional judgement was equally important to Darwin's second insight respecting "species on Galapagos archipelago".[18] In a key passage in his Ornithological Notes, written before specialists had examined his collection, Darwin recorded his suspicion that the various Galápagos mockingbirds were "only varieties" since they differed very slightly from one another and "filled the same place in Nature".[19] The phrase "only varieties" is significant in this context since naturalists traditionally used the term 'variety' to indicate groups which had been subject to some departure from type. When John Gould declared that the mockingbirds comprised three good species, Darwin could the more easily believe that the "stability of Species" had been undermined. Further, Darwin was then free to consider geographical isolation as a vehicle for species change. John Gould's recognition of a species-level distinction between the two rheas was a similar case; it invited the speculations on pages 127 and 130 of the Red Notebook. Without Gould's judgements Darwin could not have proceeded as he did, and, conversely, everything Darwin later did referred back in some way to these early professional opinions for support. Indeed in his formal presentation of his *Beagle* material Darwin took pains to emphasize that professional judgement must be relied on. Speaking of the Galápagos mockingbirds in particular he wrote:

> I may observe, that [if] some naturalists may be inclined to attribute these differences to local varieties...then the experience of all the best ornithologists must be given up, and whole genera must be blended into species.[20]

The significance of a March 1837 date for the origin of Darwin's new views on species thus derives from his reception at that time of the views of recognized zoologists with respect to key specimens from his collection. In itself this is logical enough since for Darwin to attempt an answer to the species question he had first to understand what his colleagues meant by a species in relation to his own collections.

We can now return to consider the notebook as a whole. As already mentioned, Darwin opened the notebook in May or June 1836. From internal evidence, namely the reference on page 143e to the 20 April 1837 issue of the *Athenæum* and the reference on page 178 to the subject matter of conversations Darwin was having with Robert Brown in April and May 1837, the inference can be drawn that the notebook was completed by May 1837 at the earliest. More cautiously one might wish to set the closing date at June 1837. In either case the Red Notebook was in use for something like a year. Clearly it was an important year, spanning the closing months of the voyage and the first eight or nine months back in England. It was also a year of transition, the change from one way of life to another being reflected in the pages of the notebook. Indeed, the notebook itself served partly as an instrument of adjustment to the return, for Darwin used the first part of it to plan for future publications. Scattered throughout the first part of the notebook are reminders to himself respecting his writing: 'note in Coral Paper' (page 30), 'Introduce part of the above in Patagonia paper; & part in grand discussion' (page 49), 'In Rio paper...' (page 65), 'In my Cleavage paper...' (page 101), and so on. The term 'paper' applies here to various units of Darwin's geological writings from the voyage. Darwin also used the term in that sense when he wrote to J. S. Henslow shortly after arriving home, "There is not another soul, whom I could ask, excepting yourself, to wade through & criticize [those *del*] some of those papers which I have left with you.—"[21] One paper mentioned in the Red Notebook and presumably shown to Henslow (as well as Charles Lyell) was the 'Coral Paper'. The original draft of this manuscript, written in 1835, formed the basis for Darwin's presentation on 31 May 1837 to the Geological Society of London, and, later, for the first part of the published geology from the *Beagle* voyage.[22] While the histories of the other papers referred to in the Red Notebook are not as straightforward, Darwin's intent for them was likely the same. In addition to short pieces on individual topics, Darwin also intended to write a large-scale work on the geology of South America. In 1846 Darwin realized his ambition for this 'grand discussion' of South American geology with the publication of the third part of the geology of the voyage of the *Beagle*.[23]

Entries in the Red Notebook were also directed to the furtherance of another publishing project: the *Journal of Researches*, Darwin's narrative of the 1831–1836 voyage of the H.M.S. *Beagle*.[24] While the 'Diary'[25] Darwin kept during the voyage furnished the basic narrative for his *Journal*, he included two additional kinds of material in the published work. They included references to the work of previous travellers and brief summaries of his own scientific researches. Frequently the Red Notebook was used in compiling citations of the first kind, as is evidenced by the transfer of citations from the Red Notebook to the *Journal of Researches*. Less often, but at several points most strikingly, the Red Notebook also served as an instrument for recording scientific speculations. These too passed to the *Journal* although, because of the organization of the work, rather unobtrusively.[26] In any case, it is clear on inspection that the Red Notebook served Darwin in writing the *Journal of Researches*.

The Red Notebook is thus transitional in that a number of its entries are directed towards future publications. It is also transitional in that it marks a change in the use to which Darwin put pocket-sized notebooks. While on the voyage Darwin used notebooks for recording field observations. As a result, notebook entries from the voyage are primarily observational and often not in sentence form. In contrast, most other work from the voyage, including reading notes and the finished version of daily observations, with their 'theories', 'conjectures' and 'hypotheses' (Darwin used all these terms) was written out in good sentence form on larger sized paper.[27] The Red Notebook represents a departure from this pattern, for its entries are mixed. There are still some field notes, but there are also reading notes (which are in fact sometimes notes on earlier reading notes—hence their telegraphic brevity), and, most importantly, also notes on 'theories', 'conjectures', and 'hypotheses'. After June 1837, when the Red Notebook was presumably filled, Darwin began new notebooks where the presence of theoretical inquiries became even more marked. Indeed, if one takes all of Darwin's notebooks from the *Beagle* and immediately post-*Beagle* periods together, one can see a shift from observation to theory in the notebooks, with the Red Notebook occupying a mid-way position. This change can be summarized as follows:

DESCRIPTIVE NOTEBOOKS FROM THE VOYAGE → RED NOTEBOOK → POST–VOYAGE THEORETICAL NOTEBOOKS

The change is, however, less dramatic than is suggested by this scheme, for all the notebooks show some degree of mixture in their entries; yet as an overall shift it is clear.[28] Accompanying this shift there was also a parallel change in Darwin's labelling of his notebooks. Field notebooks were labelled in a straightforward manner according to the names of the places visited. Post-*Beagle* theoretical notebooks were labelled alphabetically, presumably in deference to the abstract nature of their contents. Again the Red Notebook stood between these two groups. Its label, 'R.N.', provided no clue to its contents, and its name, the Red Notebook, merely described its physical appearance. In all likelihood the notebook went without a name until Darwin had reason to refer back to it after it was completed.[29] This would not seem to indicate any lack of regard on Darwin's part for the notebook, for indeed it was a pivotal notebook in several respects, but rather its unique standing among his notebooks.

Once the Red Notebook was filled, Darwin reorganized his method of taking notes. Where the Red Notebook contained entries on all subjects of interest, subsequent notebooks were more restricted in content. In place of the Red Notebook Darwin opened two new notebooks, one devoted to geology which he labelled 'A'.[30] At about the same time Darwin opened a second notebook, 'B', already mentioned, which he devoted to questions pertaining to the mutability of species. The generative relationship between the Red Notebook and Notebooks A and B is suggested by Figure 1.

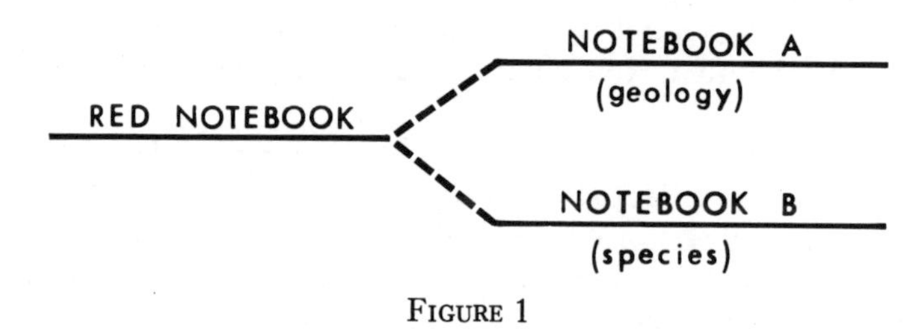

FIGURE 1

While it is beyond the scope of this introduction to describe Notebooks A and B in detail, they can be compared to the Red Notebook with respect to content. On the geological side Darwin considered a wide range of subjects in the Red Notebook. As might be expected from a notebook stimulated by field work, the largest number of geological entries in the Red Notebook pertained to specific formations and rock types. However, nearly as large in number were entries pertaining to the elevation and subsidence of the earth's crust—a subject in which Darwin was keenly interested—and what were for Darwin the attendant issues of the form of the earth—that is its shape and interior structure—and such patterns of disturbance in the earth's crust as were indicated by the occurrence of earthquakes and the presence of volcanoes and mountain chains. In addition to these major themes Darwin also made notes on other geological topics in the Red Notebook, among them the distribution of metallic veins, the preservation of fossils, erratic blocks, and life at the bottom of the sea. Equally noteworthy as the range of geological subjects considered in the Red Notebook is the enthusiasm with which they are treated, particularly those of a theoretical nature. indeed, Darwin's ambition as a theoretical geologist surfaced at several points in the notebooks; see, for example, the passage on pages 72–73 (see p. 51) beginning with the phrase, "Geology of whole world will turn out simple.—" Such passages must, of course, be read in context, and in this instance the context was provided by the Lyellian reconstruction of geology then in progress.[31]

Reflecting Darwin's enthusiasm for large theoretical issues, Notebook A follows the lead of the Red Notebook with regard to geology. For example, there is in the two notebooks a continuing interest in vertical movements of the earth's crust and an overlapping range of topics generally. Yet there are some differences between the two notebooks. Notebook A has fewer field notes than does the Red Notebook, and indeed the major piece of geological field research Darwin did in the 1837–1839 period he recorded in another notebook.[32] Also, Notebook A draws on contemporary journal literature far more than does the Red Notebook, for the obvious reason that Darwin had access to such literature only after his return to England. Another difference between the two notebooks is their relative value as documents for interpreting Darwin's geological views for the period when each was kept. In this respect the Red Notebook is the more revealing document. Yet the lesser import of Notebook A is not due to a declining interest in geology on Darwin's part during the 1837–1839 period. Indeed, in this time Darwin published seven papers on geological topics.[33] Two of these papers, that on the formation of mould and that on the 'parallel roads' of Glen Roy in Scotland, involved new field research. Further, during this same period Darwin continued working on his 'grand discussion' of South American geology, and on his studies of coral reefs and volcanic islands.[34] However, in quantity and substance Notebook A represents only a small portion, a sampler, of Darwin's geological work during the period, and for that reason it is less essential to interpreting Darwin's early geological views than is the Red Notebook.

In contrast, Notebook B and its successors represent the bulk of Darwin's theoretical work on the species question during the period when they were kept and are therefore essential in understanding his intellectual development. As is well known, Darwin did not publish his new views on species immediately upon their inception, being well aware of the generally critical attitude of his scientific colleagues towards theories asserting the mutability of species.[35] However, from the spring of 1837 on, Darwin himself was convinced of the merits of the transmutationist case and chose to pursue the subject in private without the explicit knowledge or direct support of his colleagues. In Notebook B, begun in July 1837, Darwin continued the inquiries on species begun in the Red Notebook. Once filled, Notebook B gave way to Notebooks C, D, and E, and to at least one other notebook known only from fragments.[36] By the close of Notebook C, however, Darwin's search for an explanation for adaptation had focused on the subject of behaviour, and he opened a new set of notebooks, labelled M and N, devoted in large part to the study of behaviour.[37] Like his predecessor and fellow transmutationist Jean Baptiste Lamarck, Darwin suspected that adaptive change at least in some instances might occur first in the behaviour of the organism, and Notebook M was opened with this hypothesis in mind. For the period July 1838–July 1839, Darwin was thus pursuing three related but distinguishable lines of inquiry.[38] Expressed schematically, his theoretical notebooks, which represent these lines of inquiry, developed from each other during the period from 1836–1839 as indicated in Figure 2.

It is of course possible, even probable, that other notebooks from the post-*Beagle* period await discovery and reconstruction, and that new dimensions to Darwin's work will emerge from a study of these manuscripts. Certainly within the last twenty years scholars have identified a large body of evidence which considerably illuminates the course of Darwin's labours.[39] The Red Notebook now forms part of that evidence. It stands at the beginning of that chain of events which led from Darwin's assertion of a belief in the mutability of species through his arrival at the notion of natural selection and then, after twenty years and by way of several drafts, to the publication in 1859 of the *Origin of Species*.[40] For that alone the Red Notebook is important. Moreover, apart from its place in the sequence of developments which led to the *Origin*, the Red Notebook has intrinsic merit. For example, the combination of geological, zoological, and botanical entries in the notebook is in itself important as reflecting the broad knowledge of nineteenth-century naturalists. The notebook also records much about the daily circumstances under which Darwin worked. His access to a large body of scientific and travel literature is apparent from the notebook and contradicts a common impression that he worked from only a handful of books while aboard ship. Further, the brilliance of the company Darwin kept on his return to England suggests his position, even early in his career, within one of the most influential circles of English science. Yet, and here the common view is confirmed rather than denied, Darwin's frequent mention of the names of Alexander von Humboldt and Charles Lyell substantiates his repeated references in later life to their influence on his early career. In sum the Red Notebook provides the means not only for documenting Darwin's early belief in transmutation and gauging the extent of his geological ambitions, but also for illustrating his passage from H.M.S. *Beagle* to the world of professional science.

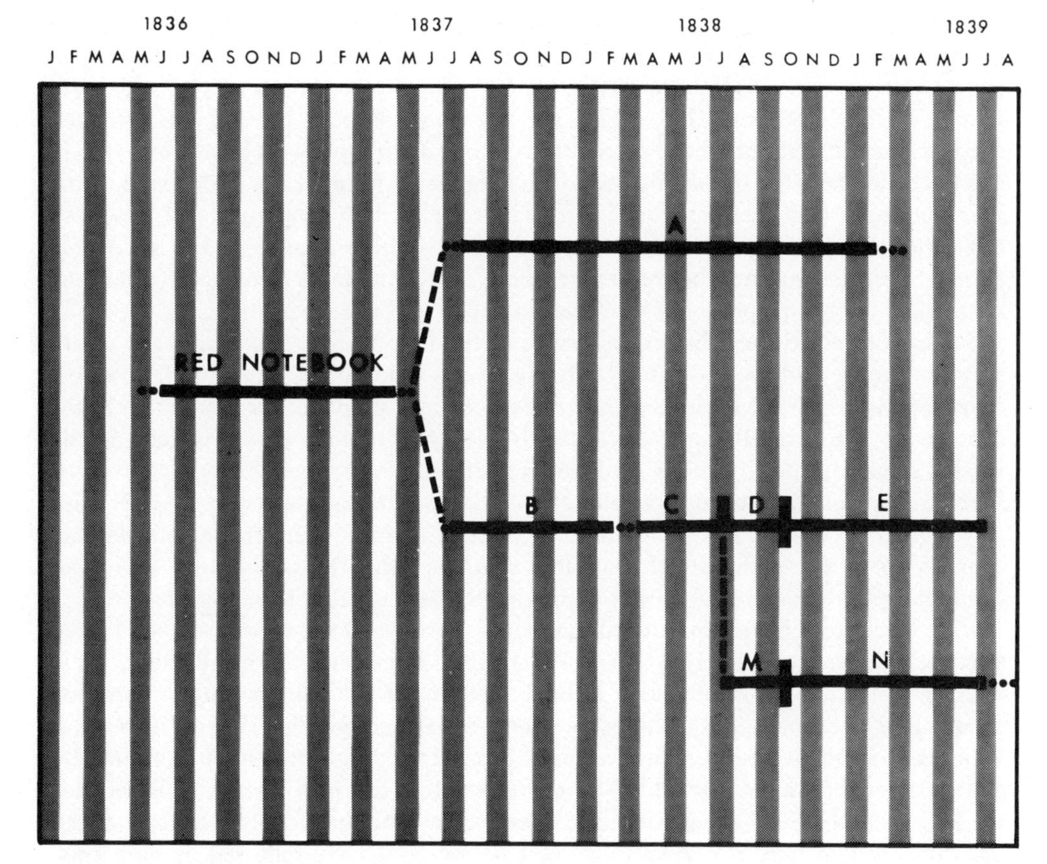

FIGURE 2. Eight Darwin notebooks kept during 1836–1839. Solid lines represent the notebooks, dotted lines uncertainties in dating, and broken lines divisions of subject matter among members of the set.

Editorial Considerations

In this edition I have intended to offer a literal transcription of the text while keeping editorial intrusions to a minimum.[41] I omitted or altered only certain obvious features of the manuscript. First, I have ignored the single-line vertical scoring which runs through most of the entries; Darwin scored his notes in this fashion when he had no further use for them. Second, I have signalled all announced changes in subject by paragraphing. In the notebook such changes are characteristically marked by a horizontal line between entries or, less frequently, by the start of a new page, rather than by paragraphing; third, except for insertions and interlineations, I have not indicated anything concerning placement of entries on the page; fourth, I have recorded the medium in which entries were written only in certain instances. In particular, up to page 113, where the original text is in pencil, I have noted those entries which are written in ink. After page 113, where the text contains both penned and pencilled entries, I have noted only those entries written in a distinctive light brown ink, which, like the penned entries in the first part of the notebook, were clearly added some time after the main text was complete. Fifth, and finally, only Darwin's text has been reproduced. The notations in other handwriting which appear in the notebook, namely the catalogue number of the notebook and some faint remains of what may have originally been a price mark, are not included in the transcription.

I have retained the other complexities of the text. Cancellations are kept where they are legible and are indicated by being enclosed in angled brackets. Those few cancellations which are omitted as illegible are mainly single letters whose identity is obscured by the cancellation mark. Darwin's alterations to the text, which include careted remarks, interlineations, and later additions, are enclosed in slanted brackets. I have not assigned dates to these alterations. Some were roughly contemporary with the original text; others, such as those in light brown ink, were made considerably later. Without surveying all of Darwin's writings from the post-*Beagle* period, it would be speculative to assign even approximate dates to all of the additions.

In spelling I have reproduced Darwin's words exactly as written where the individual letters are clearly formed. Where individual letters are indistinct, as is often the case, I have offered the probable reading of the word without comment. 'Rememb[?]g' on page 72 thus appears simply as 'remembering'. Where the reading is conjectural I have placed it within square brackets and indicated my uncertainty by a question mark. In orthography I have preserved all of Darwin's forms, except for the long 's', which I have modernized silently. In capitalization I have reproduced Darwin's usage where it is clear. Where it is unclear I have followed modern conventions. Representation of punctuation is a more serious problem, partly because nineteenth-century practice differed considerably from our own. As R. C. Stauffer has pointed out, Darwin followed a system similar to that suggested in Lindley Murray's *English Grammar*.[42] In that system commas, semicolons, colons, and periods indicate increasingly longer pauses more than they distinguish different constructions. Thus, Darwin might use a colon where a semicolon would now be employed. I have not altered his practice in this regard, which the reader should bear in mind. The representation of punctuation is also problematical since punctuation marks are by nature small and easily confused with stray marks on the page, or with pen rests. Rather than probe Darwin's intentions in this regard, I have reproduced all traditional marks of punctuation which appear within the text. Equally, I have not added punctuation where there is none. It is left to the reader, understanding the nature of the document, to tolerate inconsistencies in punctuation.

As for references I have provided footnotes on all persons named in the text. Further information on the majority of these persons can be obtained from standard biographical dictionaries. Particularly useful in this regard are the *Dictionary of National Biography* (65 vols. plus supplements; London, 1885–1900, 1901–) and the *Dictionary of Scientific Biography* (15 vols.; New York, 1970–1978). Other points I have footnoted as it seemed appropriate. I have identified or provided co-ordinates to place names only where they were not to be found in standard atlases or in Darwin's publications stemming from the voyage. In some notes I have offered cross-references to various early works of Darwin, but I have not attempted to supply a concordance between the Red Notebook and Darwin's published writings. In citing Darwin's publications I have relied primarily on R. B. Freeman, *The Works of Charles Darwin: An Annotated Bibliographical Handlist* (2nd ed.; London, 1977). Generally, in citing books in the notes, I have shortened titles; full titles are given in the bibliography. Journal titles are cited in full in both places. In the notes, unless otherwise stated, I have cited the edition of a work Darwin used if one could be determined from his citation. In many cases I was able to cite from Darwin's own copy of the work. Darwin's personal library is presently divided between Down House and the Cambridge University Library. The general catalogue of his library is H. W. Rutherford, *Catalogue of the Library of Charles Darwin now in the Botany School, Cambridge* (Cambridge, 1908). Those books presently at Cambridge University Library are described in a mimeographed pamphlet distributed by the Library entitled *Darwin Library: List of books received in the University Library Cambridge, March–May 1961.* Darwin's collection of scientific reprints is also presently housed at Cambridge University Library.

Acknowledgements

I would like to thank the present owners of Down House, the Royal College of Surgeons of England, and particularly the Secretary of the College, R. S. Johnson-Gilbert, for permission to publish the Red Notebook; the Syndics of Cambridge University Library for permission to publish excised pages from the notebook; and the Darwin family, as represented by George Pember Darwin, for their consent to the publication of the Notebook. I would also like to thank Jessie Dobson, former Curator of the Hunterian Museum, for arranging with the Royal College of Surgeons for permission to publish the Notebook.

At Down House, I am particularly indebted to the Honorary Curator of the Darwin Museum, Sir Hedley Atkins, Philip Titheradge, Custodian, and Sydney Robinson, the former Custodian. Their hospitality made working at Down House a great pleasure. For sustained assistance throughout the course of this editing project, I am also deeply indebted to Sydney Smith, Fellow of St Catharine's College, Cambridge, for sharing his rich knowledge of Darwin's writings; to Peter Gautrey, Assistant Under-librarian at Cambridge University Library, for deciphering difficult passages in the notebook and for checking my transcription of the entire notebook against the original manuscript; and to M. J. Rowlands, Librarian of the British Museum (Natural History), for encouraging and aiding the project at every point. I also wish to thank fellow Darwin scholars Frederick Burkhardt, M. J. S. Hodge, David Kohn, David Stanbury, and Frank Sulloway for specific points of information relating to the text; David Snow of the British Museum (Natural History) for supplying current scientific names for a number of Darwin's ornithological specimens; and Robert Cross, Head of Publications of the British Museum (Natural History) and Anthony P. Harvey of the British Museum (Natural History) for their aid in bringing this project to fruition. Of librarians, beyond those already mentioned, I am most indebted to the reference staff at the Library of Congress, particularly James Flatness, James Gilreath, Ann Hallstein, David Kresh, and Melissa Trevvett. I also wish to thank John Schroeder of the U.S. Geological Survey Library, Edeltraud R. Nutt, Librarian of the Geological Society of London, and M. I. Williams, Keeper of Printed Books, National Library of Wales. Photographs of the rheas which accompany this edition are from the San Diego Zoo, courtesy of Arthur Risser, those of the reconstructed *Macrauchenia patachonica* from the American Museum of Natural History, courtesy of Richard H. Tedford. Other photographs in this volume are from Cambridge University Library and the British Museum (Natural History). The diagrams in the introduction and the copies of Darwin's drawings were done by the Cartographic Service of the University of Maryland, Baltimore County. For financial support of this project, I am indebted to the National Science Foundation and the University of Maryland, Baltimore County. For encouragement and intellectual exchange, I am indebted at many levels to fellow scholars, friends, and family, of the last most particularly my parents, Emrick C. and Dorothy L. Swanson, and my husband James C. Herbert.

Notes

[1] In 1942 the Darwin family, represented by Sir Alan Barlow, husband of Nora Barlow, a grand-daughter of Charles Darwin and herself an editor of Darwin manuscripts, determined that the large collection of papers belonging to Charles Darwin in their possession would be made available to scholars. The family also determined to divide the collection, depositing the bulk of it at Cambridge University Library but reserving Charles's 'Diary' from the voyage, his *Beagle* notebooks, and some other items, particularly those relating to Down House, for the Darwin Museum.

In accordance with this arrangement twenty-four notebooks from the *Beagle* voyage are now located at Down House. These include six soft-cover notebooks, bound in two sets of three, which list specimens collected during the voyage, and eighteen hard-cover notebooks, these last having been numbered by an unknown cataloguer. Of the numbered notebooks, notebook '1' is entirely excised and bears a

London address, notebooks '2' (the Red Notebook) and '5' ('St Helena Model') are partly of post-voyage date, and the others are field notebooks from the voyage. None of the twenty-four notebooks at Down House has previously been published in its entirety, but selections from all of them are contained in Nora Barlow, ed., *Charles Darwin and the Voyage of the Beagle* (London, 1945). For information relating to the deposit of the notebooks at Down House see Darwin MSS, Cambridge University Library, vol. 156. On the founding and operation of the Darwin Museum see Sir Hedley Atkins, *Down: The Home of the Darwins* (London, 1974), chapters 8 and 9.

[2] Darwin MSS, Cambridge University Library. Divisions in this collection correspond to those in Darwin's own files. Excised pages from the Red Notebook were found in vol. 5, which contains an assortment of notes from Darwin's early life, and in vols. 40 and 42, which contain notes under several geological topics, including 'Earthquakes', 'Cleavage' and the like. A few portions of pages from the Red Notebook found at the University Library have not been included in this edition because I judged them too fragmentary to be of interest. As a rule Darwin cut out pages, or sections of pages, from his notebooks for use in future writing, placing them for reference under the appropriate heading in his files. For more information on the holdings of Darwin papers at Cambridge see the *Handlist of Darwin Papers at the University Library Cambridge* (Cambridge, 1960).

[3] Darwin and Richard Owen first met at the house of Charles Lyell on 29 October 1836. See Leonard G. Wilson, *Charles Lyell: The Years to 1841* (New Haven and London, 1972), p. 434. See also Darwin's letter to J. S. Henslow of 30 October 1836 in Nora Barlow, ed., *Darwin and Henslow: The Growth of an Idea* (Berkeley and Los Angeles, 1967), pp. 118–119.

[4] The earliest known reference to Richard Owen's identifications of Darwin's South American fossil mammals occurs in a letter from Owen to Charles Lyell dated 23 January 1837. The letter is reproduced in full in Leonard G. Wilson, *Charles Lyell: The Years to 1841* (New Haven and London, 1972), pp. 436–437. For Lyell's use of this information see Charles Lyell, 'Presidential Address to the Geological Society of London [17 February 1837]', *Proceedings of the Geological Society of London*, vol. 2 (1838), pp. 510–511. In reporting Owen's identifications Lyell summarized the results for science of the new specimens (p. 511): "These fossils...establish the fact that the peculiar type of organization which is now characteristic of the South American mammalia has been developed on that continent for a long period, sufficient at least to allow of the extinction of many large species of quadrupeds. The family of the armadillos is now exclusively confined to South America and here we have from the same country the Megatherium, and two other gigantic representatives of the same family. So in the Camelidæ, South America is the sole province where the genus Auchenia or Llama occurs in a living state, and now a much larger extinct species of Llama is discovered. Lastly, among the rodents, the largest in stature now living is the

Capybara, which frequents the rivers and swamps of South America and is of the size of a hog. Mr. Darwin now brings home from the same continent the bones of a fossil rodent not inferior in dimensions to the rhinoceros. These facts elucidate a general law previously deduced from the relations ascertained to exist between the recent and extinct quadrupeds of Australia; for you are aware that to the westward of Sydney on the Macquarie River, the bones of a large fossil kangaroo and other lost marsupial species have been met with in the ossiferous breccias of caves and fissures."

[5] The notes I refer to are contained in volumes 30 and 31 of the Darwin manuscript collection at Cambridge University Library.

[6] Darwin's earliest recollection of the origin of his new views occurs in a journal entry for the year 1837 and reads as follows: "In July opened first note book on 'transmutation of Species'—Had been greatly struck from about month of previous March on character of S. American fossils—& species on Galapagos Archipelago. These facts origin (especially latter) of all my views." This entry appears in a notebook begun in August 1838. A fair copy of the notebook was published as Sir Gavin de Beer, ed., 'Darwin's Journal', *Bulletin of the British Museum* (*Natural History*) Historical Series, vol. 2 (1959) where the quoted entry appears on page 7. The entry as quoted here is taken from the original notebook which has come to light since 1959.

[7] Darwin's full account of his conversion to transmutationist views as it appears in his autobiography written some forty years after the events reads as follows: "During the voyage of the *Beagle* I had been deeply impressed by discovering in the Pampean formation great fossil animals covered with armour like that on the existing armadillos; secondly, by the manner in which closely allied animals replace one another in proceeding southwards over the Continent; and thirdly, by the South American character of most of the productions of the Galapagos archipelago, and more especially by the manner in which they differ slightly on each island of the group; none of these islands appearing to be very ancient in a geological sense. It was evident that such facts as these, as well as many others, could be explained on the supposition that species gradually become modified; and the subject haunted me." The passage is taken from Nora Barlow, ed., *The Autobiography of Charles Darwin* (London, 1958), pp. 118–119.

[8] I refer here to Jean Baptiste de Lamarck (1744–1829) whose views Darwin encountered most forcefully although critically expressed in the second volume of Charles Lyell's *Principles of Geology* (1832).

[9] The most striking instance in Darwin's collection where professional judgement recognized small differences as indicating good species was the Galápagos mockingbirds. See note 159 to the text and subsequent discussion in this introduction.

[10] See notes 174 and 234 to the text of the Red Notebook.

[11] See note 6 to this introduction.

[12] In his later account Darwin recalled being particularly impressed by the similarity between "great fossil animals covered with armour" [glyptodonts] and living armadillos. See Nora Barlow, ed., *The Autobiography of Charles Darwin* (London, 1958), p. 118. Richard Owen confirmed Darwin's interpretation of the affinity of the 'glyptodont' and the armadillo in January 1837. See Leonard Wilson, *Charles Lyell: The Years to 1841* (New Haven and London, 1972), p. 437.

[13] Red Notebook, p. 130 and note 159. It should be stressed that the relationship of the South American and Galápagos mockingbirds is exactly that which Darwin described in his autobiography: "During the voyage of the *Beagle* I had been deeply impressed...by the South American character of most of the productions of the Galapagos archipelago, and more especially by the manner in which they differ slightly on each island of the group; none of these islands appearing to be very ancient in a geological sense." Nora Barlow, ed., *The Autobiography of Charles Darwin* (London, 1958), p. 118.

[14] Sir Gavin de Beer, ed., 'Darwin's Journal', *Bulletin of the British Museum* (*Natural History*) Historical Series, vol. 2 (1959), p. 7, refers to the opening of the first notebook on 'transmutation of Species' in July 1837. This is Notebook B. Darwin filled this notebook sometime in February or March 1838. The exact date is in doubt. Darwin referred in his heading to the notebook (probably added in 1844 when he was arranging his papers) that he completed the notebook at the beginning of February. In fact the notebook must have been completed somewhat later, for p. 235 refers to the issue of the *Athenæum* of 24 February 1838. Since the notebook ran to another 29 pages of text after p. 235, it was probably completed no earlier than the end of the month. Notebook B is numbered as vol. 121 in the Darwin MSS, Cambridge University Library, and is published as: Sir Gavin de Beer, ed., 'Darwin's Notebooks on Transmutation of Species. Part I', First Notebook (July 1837–February 1838), *Bulletin of the British Museum* (*Natural History*) Historical Series, vol. 2 (1960), pp. 23–73. A number of the excised pages to Notebook B, and to its successors, Notebooks C, D, and E, were later published as: Sir Gavin de Beer and M. J. Rowlands, 'Darwin's Notebooks on Transmutation of Species. Addenda and Corrigenda', *Bulletin of the British Museum* (*Natural History*) Historical Series, vol. 2 (1961), pp.185–200, and Sir Gavin de Beer, M. J. Rowlands, and B. M. Skramovsky, 'Darwin's Notebooks on Transmutation of Species, Part VI, Pages Excised by Darwin', *Bulletin of the British Museum* (*Natural History*) Historical Series, vol. 3 (1967), pp. 129–176.

[15] There is, incidentally, another passage in Notebook B which supports the more general conclusion that Notebook B was the successor to the Red Notebook. On page 153e of Notebook B Darwin referred to the Red Notebook as follows: "See R.N. p. 130 speculations range of allied species. p. 127 p. 132. There is no more wonder in extinction of individuals than of species." Clearly the Red Notebook, at least to page 130, was already in existence by the time Darwin made this entry in Notebook B.

[16] The dates when important specimens referred to in the Red Notebook were identified by professional zoologists are as follows: the *Macrauchenia* was referred to descriptively, although not by that name, in a letter written to Richard Owen dated 23 January 1837; the Galápagos mockingbirds were described at a meeting of the Zoological Society of London on 28 February 1837; and the new species of South American rhea was described at a meeting of the Zoological Society of London on 14 March 1837. See also notes 152, 159, and 149 to the text.

[17] Nora Barlow, ed., *The Autobiography of Charles Darwin* (London, 1958), p. 118. The similarity of the polygonal plates of the glyptodon specimen to those of the armadillo was noticed by Darwin "immediately I saw them." (Darwin to J. S. Henslow, 24 November 1832, in Nora Barlow, ed., *Darwin and Henslow: The Growth of An Idea* [Berkeley and Los Angeles, 1967], p. 61.) Darwin was not the first to note the similarity of the large plates to the smaller ones of the armadillo. On this point see Thomas Falkner, *A Description of Patagonia* (London, 1774), p. 55.

[18] Sir Gavin de Beer, ed., 'Darwin's Journal', *Bulletin of the British Museum* (*Natural History*) Historical Series, vol. 2 (1959), p. 7.

[19] Nora Barlow, ed., 'Darwin's Ornithological Notes', *Bulletin of the British Museum* (*Natural History*) Historical Series, vol. 2 (1963), p. 262. The full paragraph reads as follows: "I have specimens from four of the larger islands; the two above enumerated, and (3349: female. Albermarle Isd.) & (3350: male: James Isd).—The specimens from Chatham & Albermarle Isd appear to be the same; but the other two are different. In each Isld. each kind is *exclusively* found: habits of all are indistinguishable. When I recollect, the fact that the form of the body, shape of scales & general size, the Spaniards can at once pronounce, from which Island any Tortoise may have been brought. When I see these Islands in sight of each other, & [but *del.*] possessed of but a scanty stock of animals, tenanted by these birds, but slightly differing in structure & filling the same place in Nature, I must suspect they are only varieties. The only fact of a similar kind of which I am aware, is the constant asserted difference — between the wolf-like Fox of East & West Falkland Islds. — If there is the slightest foundation for these remarks the zoology of Archipelagoes—will be well worth examining; for such facts [would *inserted*] undermine the stability of Species." For more extended discussion of this passage see Sandra Herbert, 'The Place of Man in the Development of Darwin's Theory of Transmutation, Part I. To July 1837', *Journal of the History of Biology*, vol. 7 (1974), pp. 236–240.

[20] From Darwin's commentary on specimens in John Gould, *The Zoology of the Voyage of H.M.S. Beagle. Part III: Birds.* 5 numbers. (London, 1838–1841), pp. 63–64.

[21] Darwin to J. S. Henslow, 30 October 1836, in Nora Barlow, ed., *Darwin and Henslow: The Growth of An Idea* (Berkeley and Los Angeles, 1967), p. 122.

[22] The original draft and a fair copy of the paper are contained in vol. 41 of the Darwin manuscript collection at Cambridge University Library. A transcription of the original draft is available in print as 'Coral Islands by Charles Darwin', Introduction, Map and Remarks by D. R. Stoddart, *Atoll Research Bulletin*, No. 88 (1962). Darwin's theory of coral island formation first appeared in print as Charles Darwin, 'On Certain Areas of Elevation and Subsidence in the Pacific and Indian Oceans, As Deduced from the Study of Coral Formations', *Proceedings of the Geological Society of London*, vol. 2 (1838), pp. 552–554. A fuller version of the theory appeared later as Charles Darwin, *The Structure and Distribution of Coral Reefs. Being the First Part of Geology of the Voyage of the Beagle, under the Command of Capt. Fitzroy, R.N. during the Years 1832–1836* (London, 1842).

[23] Charles Darwin, *Geological Observations on South America. Being the Third Part of the Geology of the Voyage of the Beagle, under the Command of Capt. Fitzroy, R.N. during the Years 1832–1836* (London, 1846).

[24] Charles Darwin, *Journal of Researches into the Geology and Natural History of the Various Countries Visited by H.M.S. Beagle* (London, 1839). Also published as volume 3 of Robert Fitzroy, ed., *Narrative of the Surveying Voyages of His Majesty's Ships Adventure and Beagle...1832–1836* (London, 1839).

[25] See Nora Barlow, ed., *Charles Darwin's Diary of the Voyage of H.M.S. Beagle* (Cambridge, 1933).

[26] Compare, for example, the Red Notebook (page 132) and the *Journal of Researches* (page 262) on 'associated life' or the Red Notebook (pages 129–130, 132–133) and the *Journal of Researches* (pages 208–212) on the successions of fossil and living forms and the causes of extinction of species. The references to the *Journal* are from Charles Darwin, *Journal of Researches into the Geology and Natural History of the Various Countries Visited by H.M.S. Beagle* (London, 1839). It is worth remarking that the presence of theoretical passages in the *Journal of Researches* is disguised by the organization of the work, which is chronological rather than logical, and by the lack of an adequate index.

[27] The primary series of formal notes from the voyage are contained in vols. 30–31 (zoology) and vols. 32–38 (geology) of the Darwin MSS at Cambridge University Library.

[28] There is one notable exception to this schema: the field notebook Darwin carried with him when he investigated the 'parallel roads' of Glen Roy in Scotland in June and early July 1838. It is almost entirely observational in character. The notebook is numbered as vol. 130 in the Darwin MSS at Cambridge University Library.

[29] In other writing Darwin usually referred to the notebook as 'R.N.' or even the 'R.N. notebook' rather than as the 'Red Notebook'. I am grateful to M. J. S. Hodge

of the University of Leeds for locating and informing me of an instance where Darwin used the full name of the notebook. It occurs on the verso of a scrap of paper in vol. 29 (iii) of the Darwin MSS at Cambridge University Library. I have since come upon other instances in Darwin's notes where he used the full name of the notebook, but they are rare.

[30] Notebook A was kept from about July 1837 to the late spring of 1839. Both dates are conjecture on my part. The first datable reference of relevance in the notebook occurs on p. 15e and is to the August 1837 issue of *L'Institut*. (Since Notebook A was filled fairly evenly and slowly, overall at the rate of fewer than ten pages a month, the August date on p. 15e would not preclude an earlier opening date than July for the notebook.) With respect to the closing date, nothing is certain, but Darwin was already abstracting from the notebook as early as 24 February 1839. In a sense the notebook had served its purpose by that date. Notebook A, which I am presently editing for publication, is numbered as vol. 127 in the Darwin MSS at Cambridge University Library.

[31] While no complete account of Darwin's geological work exists, the subject of Lyell's influence on Darwin during the post-voyage period is treated in Leonard G. Wilson, *Charles Lyell: The Years to 1841* (New Haven and London, 1972), chapter 7. For a brief account of Lyell's influence on Darwin with respect to one particular problem see: Martin Rudwick, 'Darwin and Glen Roy: A "Great Failure" in Scientific Method?', *Studies in the History and Philosophy of Science*, vol. 5 (1974), pp. 165–167.

[32] Darwin's major piece of field research in the 1837–1839 period was his investigation of the so-called 'parallel roads' of Glen Roy. He recorded his observations from this research in the notebook described in note 28 above.

[33] Darwin published the following papers on geological topics during the years 1837–1839: 'Observations of Proofs of Recent Elevation on the Coast of Chili, Made during the Survey of His Majesty's Ship Beagle, Commanded by Capt. Fitzroy, R.N.', [Read 4 January 1837] *Proceedings of the Geological Society of London*, vol. 2 (1838), pp. 446–449; 'A Sketch of the Deposits Containing Extinct Mammalia in the Neighbourhood of the Plata', [Read 3 May 1837] *Proceedings of the Geological Society of London*, vol. 2 (1838), pp. 542–544; 'On Certain Areas of Elevation and Subsidence in the Pacific and Indian Oceans, as Deduced from the Study of Coral Formations', [Read 31 May 1837] *Proceedings of the Geological Society of London*, vol. 2 (1838), pp. 552–554; 'On the Formation of Mould', [Read 1 November 1837] *Transactions of the Geological Society of London*, 2nd ser., vol. 5, pt. 3 (1840), pp. 505–509; 'On the Connexion of Certain Volcanic Phenomena in South America; and on the Formation of Mountain Chains and Volcanos, as the Effect of the Same Power by which Continents Are Elevated', [Read 7 March 1838] *Transactions of the Geological Society of London*, 2nd ser., vol. 5, pt. 3 (1840), pp. 601–631; 'Observations on the Parallel

Roads of Glen Roy, and of Other Parts of Lochaber in Scotland, with an Attempt to Prove that They Are of Marine Origin', [Read 7 February 1839] *Philosophical Transactions of the Royal Society of London*, vol. 129 (1839), pp. 39–81; and 'Note on a Rock Seen on an Iceberg in 61° South Latitude', *Journal of the Royal Geographical Society of London*, vol. 9 (1839), pp. 528–529. These citations are from Paul H. Barrett, ed., *The Collected Papers of Charles Darwin* (Chicago and London, 1977), vol. 1, pp. v–vi and 41–139.

[34] The three parts of Darwin's geological results from the *Beagle* voyage as published in book form were as follows: *The Structure and Distribution of Coral Reefs* (London, 1842); *Geological Observations on the Volcanic Islands Visited during the Voyage of H.M.S. Beagle* (London, 1844); and *Geological Observations on South America* (London, 1846). Although published over five years, Darwin regarded the three parts as forming a single work.

[35] Darwin's first public announcement of his theory of evolution through natural selection was at a meeting of the Linnean Society held on 1 July 1858. For an assessment of the meeting see J. W. T. Moody, 'The reading of the Darwin and Wallace papers: an historical "non-event"', *Journal of the Society for the Bibliography of Natural History*, vol. 5 (1971), p. 474–476. The resultant publication with Alfred Russel Wallace appeared under a general title as: 'On the Tendency of Species to Form Varieties; and on the Perpetuation of Varieties and Species by Natural Means of Selection', *Journal of the Proceedings of the Linnean Society of London, Zoology*, vol. 3 (1858), pp. 45–62. The next year saw the publication of Darwin's *On the Origin of Species by Means of Natural Selection* (London, 1859). On the relation of Darwin's theoretical work to his pattern of publication see Sandra Herbert, 'The Place of Man in the Development of Darwin's Theory of Transmutation, Part II', *Journal of the History of Biology*, vol. 10 (1977), pp. 157–196.

[36] Notebook B was begun in July 1837 and completed in February or March 1838. Notebook C was begun about March 1838. When completed, it was replaced by Notebook D, opened on 15 July 1838. When filled, Notebook D was in turn replaced by Notebook E, begun about 2 October 1838 and ended on 10 July 1839. Notebooks B to E comprise vols. 121–124 of the Darwin MSS at Cambridge University Library. They have appeared in print as Sir Gavin de Beer, ed., 'Darwin's Notebooks on Transmutation of Species. Parts I–IV', Sir Gavin de Beer and M. J. Rowlands, eds., 'Darwin's Notebooks on Transmutation of Species. Addenda and Corrigenda', and Sir Gavin de Beer, M. J. Rowlands, and B. M. Skramovsky, eds., 'Darwin's Notebooks on Transmutation of Species. Part VI. Pages Excised by Darwin', *Bulletin of the British Museum (Natural History)* Historical Series, vol. 2 (1960–1961), pp. 23–200; vol. 3 (1967), pp. 129–176. The unexcised portions of Notebooks B, C, D, and E correspond to Parts I–IV in this series. In addition to Notebooks B–E, twenty-two pages of another notebook on transmutation have been located among the Darwin

manuscripts at Cambridge University Library. This 'Torn-up Notebook' was first assembled by Sydney Smith and announced in his Sandars Lectures of 1966–1967. The notebook was opened in about July 1839 and, on the evidence of a dated page, was kept through June 1841. A plausible closing date for the notebook might be autumn 1841. This notebook is presently being edited for publication by Sydney Smith and David Kohn. From the collections at the Cambridge University Library, David Kohn has also located another notebook from the early period of Darwin's work on species. The six extant pages of this notebook date from the summer of 1842 and pertain to the subject of the cross-fertilization of flowers.

[37] Notebook M was opened on 15 July 1838. When filled it was replaced by Notebook N, opened on 2 October 1838. Entries in Notebook N declined by mid-summer 1839, though occasional entries were made as late as 1840. Notebooks M and N comprise vols. 125 and 126 of the Darwin MSS at Cambridge University Library. They were edited by Paul H. Barrett and appear in Howard E. Gruber and Paul H. Barrett, *Darwin on Man: A Psychological Study of Scientific Creativity* (New York, 1974).

[38] In September 1838 Darwin was in fact pursuing four separable lines of inquiry as indicated by his method of note-taking. The fourth line of inquiry was 'generation' meaning, loosely, reproduction. Generation was an important topic throughout Notebooks B–E but one given especial prominence in Notebook D by virtue of the fact that the end portion of the notebook (pages 152–180) was set aside for it alone. Darwin opened this section of the notebook on 11 September 1838; presumably it was filled by 2 October 1838 as the heading of the notebook indicates.

[39] For a review of work since 1959 see John C. Greene, 'Reflections on the Progress of Darwin Studies', *Journal of the History of Biology*, vol. 8 (1975), pp. 243–273.

[40] The Red Notebook contains the earliest known evidence of Darwin's belief in the mutability of species. From his own later accounts Darwin's arrival at the notion of natural selection was precipitated by reading Thomas Robert Malthus' *An Essay on the Principles of Population.* Darwin recorded his reading of Malthus in an entry in Notebook D dated 28 September 1838; see Sir Gavin de Beer, M. J. Rowlands, and B. M. Skramovsky, eds., 'Darwin's Notebooks on Transmutation. Part VI. Pages Excised by Darwin', *Bulletin of the British Museum* (*Natural History*) Historical Series, vol. 3 (1967), pp. 162–163. Darwin's earliest draft of his theory was his 'Sketch' of 1842, followed by his lengthier *Essay* of 1844. For these see Charles Darwin and Alfred Russel Wallace, *Evolution by Natural Selection* (with a foreword by Sir Gavin de Beer) (Cambridge, 1958). In 1856 Darwin began his longest exposition of his argument. For the reconstructed text of this version see R. C. Stauffer, ed., *Charles Darwin's Natural Selection* (Cambridge, 1975). The theory finally came before the public in 1858 in a brief announcement to the Linnean Society of London (see

note 35, above) and then, a year later, in the form in which it is generally known: Charles Darwin, *On the Origin of Species by Means of Natural Selection* (London, 1859).

[41] In setting an editorial standard I have made extensive use of the work of other Darwin editors, particularly: P. Thomas Carroll, *An Annotated Calendar of the Letters of Charles Darwin in the Library of the American Philosophical Society* (Wilmington, Delaware, 1976), pp. xxvii–xxxvii; Sir Gavin de Beer, ed., 'Darwin's Notebooks on Transmutation of Species. Part IV', *Bulletin of the British Museum (Natural History)* Historical Series, vol. 2 (1960), pp. 158–159; Howard E. Gruber and Paul H. Barrett, *Darwin on Man: A Psychological Study of Scientific Creativity* (New York, 1974), pp. xviii–xxii; and R. C. Stauffer, *Charles Darwin's Natural Selection* (Cambridge, 1975), pp. ix, 15–21. The 'Style Manual and Guide to Editorial Practice' governing the future publication of *The Collected Letters of Charles Darwin*, jointly edited by Frederick Burkhardt, Sydney Smith, David Kohn, and William Montgomery, has also been consulted. In addition I found helpful G. Thomas Tansell, 'The Editing of Historical Documents', *Studies in Bibliography* [Papers of the University of Virginia Bibliographical Society], vol. 31 (1978), pp. 1–56.

[42] R. C. Stauffer, ed., *Charles Darwin's Natural Selection* (Cambridge, 1975), pp. 20–21.

front
cover

R.N.

The front cover of the Red Notebook, labelled 'R.N.'

inside front cover

up to [1° / *or* 1st of ?] July 1835. the excess of harbor = 180
See Daubisson both Volumes,[1] and Molina 1st Vol[2] ⌈& Lyell⌉[3]
Sailed, [27th ?]
⟨Friday, gale 29th⟩
Friday
Thursday 29th gale
⌈Lyell's Geology⌉[4]

⌈The living atoms having definite existence, those that have undergone the greatest number of changes towards perfection (namely mammalia) must have a shorter duration, than the more constant: This view supposes the simplest infusoria same since commencement of world.—⌉[5]|

⌈ ⌉	Darwin's addition
⟨ ⟩	Darwin's cancellation
[ ]	Editor's remark
[. . .?]	Uncertain reading
|	End of notebook page
e	Wholly or partly excised page

1–4e [not located] |

5e La. billardiere mentions the floating marine confervæ, is very common within E. Indian Archipelago, no minute description, calls it a Fucus. P [Vol I 287][6]

P 379. Henslow Anglesea, nodules in Clay Slate. major axis 2.1/2 ft. — singular structure of nodule, constitution [same as] of slate same. — longer axis in line of Cleavage. laminæ fold round them;[7] Quote this. Valparaiso Granitic nodules in Gneiss. |

6e Epidote seems commonly to occur where rocks have undergone action of heat. it is so found in Anglesea, amongst the varying & dubious granites. — Wide limits of this mineral in Australia. Fitton's appendix[8]

Would Slate. & unstratified rocks show any difference in facility of conducting Electricity? Would minute particles have a tendency to change their position? |

7e Carbonate of Lime disseminated through the great Plas Newydd dike. — Mem tres Montes. ((Henslow Anglesea))[9]

great variety in nature of a dike. — Mem. at Chonos & Concepcion. P. 417[10]

Veins of quartz exceedingly rare Mem C. [Cape] Turn P. 434 & 419[11]

As Limestone passes into schist scales of chlorites — Mem. Maldonado P 375[12]

Much Chlorite in some of the dikes. — P 432.[13] as in Andes. |

8e In Dampier's voyage there is a mine of metereology with respect to the discussion of winds & storms:[14] — [in Volney's travels also][15]

Dampier's last voyage to New Holland P 127. — Caught a shark 11 ft long.[16] "Its maw was like a leathern sack, very thick & so tough that a sharp knife could not cut it: in which we found the Head & Boans of a Hippotomus; the hairy lips of which were still sound and not petrified, and the |

9 jaw was also firm, out of which we pluckt a great many teeth, 2 of them, 8 inches long, & as big as a mans thumb, the rest not above half so long; The maw was full of jelly which stank extreamly." — This shark was caught in Shark's Bay. Lat 25°.[17] The nearest of the E Indian Islands. namely Java is 1000 miles distant! Where are Hippotami found in that Archipelago? Such have never been observed in Australia |

10 Dampier also repeatedly talks about the immense quantities of Cuttle fish bones floating on the surface of the ocean, before arriving at the Abrolhos shoals. — [18]

N.B. The view of the Volcanos of the chain of the Cordilleras as arising from [the expulsion of fluid nucleus through] faults or fissures, produced by the elevations of those mountains on the continent of S. America is inadmissible [may have happened from incipient elevation.] The volcanos originated |

11 in the bottom of the ocean. & the present Volcanos have been said to be merely accidental apertures still open. — The fault like appearance [arising from the manner of horizontal upheaval] of the shore of the Pacifick is 60 miles distant from the grand ancient volcanic axis [of the Andes]. — [Has this fault determined side of volcanic activity.] That axis was produced, from a fissure in a deep & therefore weak part of the ocean's bottom. |

12 With respect to Sharks distributing fossil remains: Sharks followed Capt. Henry's vessel from the Friendly Isles. to Sydney; know by having been seen & from the contents of its maw, amongst which were things pitched over board early in the passage!! — [19]

M. Labillardiere in Bay of Legrand, (SW part). describes a Small granite Is$^{d.}$ capped by Calcareous rock;[20] following |

13– [not located]

14e

15e Find instances; The whole coast of New Holland shoals much: Dampier remarks on great flats on the NW coast: —[21] 8 leagues, from Sydney 90 fathoms La Peyrouse.[22]

South of Mocha; 19 miles. 65 Fathoms

Vide facts in Beechey. on NW coast of America[23]

off Cape of Good Hope 70 fathoms 20 miles from the shore? Beagle

Coast of Brazil? where not rivers [in my Coral paper][24] |

16e

	leagues		Fathoms
Parallel of St Catherine [27° 30′ S.][25]	18	—	70
Paranagua [25° 42′ S.]	12	—	40
St Sebastian [23° 52′ S.]	12		50
Joatingua SE [23° 22′ S.]	5		35
R. de Janeiro SE [23° 58′ S.]	18		77
C. Frio [23° S.]	7		60

Soundings about same as last to N. of C. Frio Except at Abrolhos. [18° S.]

Bahia [12° 57′ S.]	8		$\dot{\overline{200}}$
Morro S. Paulo [13° 22′ S.]	9		$\dot{\overline{120}}$
Garcia de Avila [lighthouse] [12° 35′ S.]	9		124
Itapicuru [R.] [11° 46′ S.]	9		200
R. Real [11° 31′ S.] & [R.] Sergipe [11° 10′ S.]	20		190
R. San Francisco [10° 32′ S.]	10		50
Whole coast to Olinda [8° S.]	9–10	=	30–40

at twice or [18–20][26] ⟨60⟩ — 80 $\dot{\overline{120}}$ parallel of Olinda

Shoaler N. of Olinda. — a little WNW of C. Rock. [5° 29′ S.] still shoaler, coast composed of sand dunes. 15 — 15

Does not seem to consider this a very shoal coast.[27]

Beyond the 10 or 12 leagues sea deepens suddenly. coast of Brazil generally. — |

17 M[rs] Power at Port Louis talked of the <u>extraordinary</u> freshness of the streams of Lava in Ascencion known to be inactive 300 years?[28]

No Volcanic Earthquakes or Hot Springs in T. del Fuego = The Wager's Earthquake the most Southern one I have heard of[29] |

18 In a preface, it might be well to urge, geologists to compare whole history of Europe, with America; I might add I have drawn all my illustrations from America, purposely to show what facts can be supported from that part of the globe: & when we see conclusions substantiated over S. America & Europe. we may believe them applicable to the world. — |

19e My general opinion from the examination of soundings, from about 80 fathoms & upwards. that life is exceedingly rare, at the bottom of the sea. —⸢certainly data insufficient, yet good⸣ ⸢(I suspect fragments of shells will generally be found to be old & dead)⸣ ⸢(I have not kept a record)⸣ In looking over the lists of organic remains in De la Beche,[30] for the older formations I must believe they ⸢the limestones⸣ have been formed in shallow water: so have the Conglomerates: Yet this view is directly opposed to common opinion |

20e The Tertiary formation South of the Maypo at one period of elevation must in its configuration have resembled Chiloe

In De La Beche, article "Erratic blocks" not sufficient distinction is given to angular & rounded. — [31]

Fox Philosoph. Transactions on metallic veins. 1830 P. 399. — [32] Carne. Geolog. Trans: Cornwall ⸢Vol II⸣[33] |

21 It is a fact worth noticing that cryst of glassy felspar in Phonolite arrange themselves in determinate planes ∴ such action can take place in melted rocks

The frequent coincidence of line of veins & cleavage is important; veins appearing a galvanic phenomenon, so probably will the Cleavage be

There is a resemblance at Hobart town between the older strata & the bottom of sea near T. del Fuego. — |

22 Is there account of Baron Roussin's voyage. — [34]

In Europe proofs of many oscillations of level, which in the nature of strata & Organic remains does not appear to have taken place in the Cordillera of S. America.

Study Geolog: Map of Europe

Conybeare. Introduct XII P. silicified bones not common in Britain. Mem Concepcion Says Echinites. Encrinites. Asteriæ, usually petrified into |

23 a peculiar cream-coloured Limestone: [35] the strange substitution of matter in shells, like Concretions & laminæ show what movements take place in semiconsolidated rocks

P xv. mentions in what formations Conglomerates are found. — [36]

The above oscillations remarkable because the formations are now seen in regular descending steps |

24 Mem.; rapidity of germination in young corals. — vide L. Jackson's paper. Philosoph Transact: [37] at R. de Janeiro. Coquimbo. Balanidæ. at Concepcion.

Humb: Pers. N. vii P. 56 [38] Serpentine form: of Cuba for comparison (?) with St Pauls |

25–26e [not located]

27 The frequency of shells in the Calc. Sandstone Concret, is connected with frequency of shells in flints in Chalk

New Providence more hilly than others of the Bahama consists of rock & sand mixed with sea shells — about 500 Is$^{d.}$ & great banks. effect of Elevation. United service Journal[39] |

28 In the Iron sand formation ⟨would⟩ wood converted into siliceous pyritous & coaly matter. Mem: Chiloe

In the endless cycle of revolutions. by actions of rivers currents. & sea beaches. All mineral masses must have a tendency. to mingle; The sea would separate quartzose sand from the finer matter resulting from degradation of Felspar & other minerals containing

Alumen. — This matter |

29 accumulating in deep seas forms slates: How is the Lime separated; is it washed from the solid rock by the actions of Springs or more probably by some unknown Volcanic process? How does it come that all Lime is not accumulated in the Tropical oceans detained by Organic powers. We know |

30 the waters of the ocean all are mingled. These reflections might be introduced either in note in Coral Paper or hypothetical origin of some sandstones, as in Australia. — Have Limestones all been dissolved. if so sea would separate them from indissoluble rocks? Has Chalk |

31 ever been dissolved?

Singularity of fresh water at Iquiqui. not from rain, because alluvium saline; Mem: on coast of Northern Chili as springs become rarer, so does the rain, therefore such rain is cause, hence at least no water is absorbed into the earth

⟨I did not see one dike in the whole Galapagos Arch; because no sections, same cause as no [colour?]⟩|

32 Sir J. Herschels idea of escape of Heat prevented by sedimentary rocks, & hence Volcanic action, contradicted by Cordillera, where that action commenced before any great accumulation of such matter. —[40]

Dr A. Smith says. that Boulders do not occur in the South African plains. —[41] Sydney no |

33e I believe the secondary? formations of Brazil, all originate from the decomposition of Granitic rocks Mem. Chanticleers voyage at ⟨[J?] [Maranh?]⟩ Pernambuco.[42]

[the following is a newspaper clipping pasted on the page:[43] EARTHQUAKE AT SEA. — Extract from the log-book of the *James Cruikshank*, Captain John Young, on her voyage from Demerara to London: — "Feb. 12, 1835. At 10h. 15m. a severe shock of earthquake shook the ship in a most violent manner. Although it lasted about a minute, there was no uncommon ripple on the water. It was quite calm at the time. Latitude 8 deg. 47 min. N: longitude 61 deg. 22 min. W. mid. calm and clear.] Caermarthen Journal |

34e I look at the cessation northwards of the Coal in Chili as clearly bearing a relation to present position of ⟨Coal⟩ Forests. These thick beds of Lignite stratified with substances so like the Coal measures in England (Excepting Conglomerates?) [& absence of limestone?] have been collected on the open coast. Perhaps as at Concepcion. favoured by basin formed by outlying rocks; (such as between Mocha & main land). At Carelmapu. — Within Chiloe: — |

35e On open coast, near where Challenger was lost:[44] I know no reason for supposing these matters are not now collecting, in the bottom of an open & not deep sea. — (Character of coast regular & ⟨not very⟩ rather deep soundings, 60–100 fathoms 2 & 3 miles from shore. V. Chart) Every winter torrents must bring much vegetable matter from thickly wooded mountains, probably chiefly leaves. — This position agrees with character of.. [in Basins from rivers. & natural position] |

36e position at N.S. Wales & Van Diemen's land. —

Whole coast S. of Concepcion where there are Tertiary strata there is Coal — ? | No | shells in all cases. [.Mytilus. —][45]

[at Guacho] [on N. Chile? Washington. —][46]

Mem: Micaceous formation of Chonos. interesting from great quantity of altered Carbonaceous shales

Examine chart of Patagonian coast to see proportional cliff & low or sloping land

What are the "palatal Tritores" found in the coraliferous mountain Limestone |

37e are they allied to the jaws of the Cocos fish

Rio Shells argument for rise

In Cordillera, the dikes do not generally appear to have fallen into lines of faults

I do not think so many faults in Cordillera, as in English Coal field — because lowered & raised — so on — but gradually & simply raised

No Faults in Patagonia[,] enormous extent; if lowered again & covered no sign of upheaval |

38e To Cleavage add other instances in old world of symetrical structure. East India Archipelago. [Aleutian Arch. —] V. Fitton. Australia:[47] cases in Europe. —

Auvergne. very little Pumice, though Trachyte. same fact in Galapagos. Daubeny P 24[48]

[V. back of page 1 of New Zealand Geological Notes.][49]

at St. Helena. This structure was very clear at base of great lava cliffs[50] [Fig. 1]

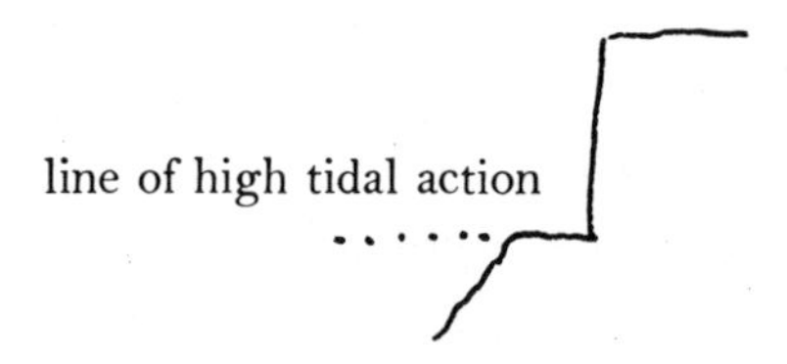

NB. patches of modern Conglomerates [Fig. 2]

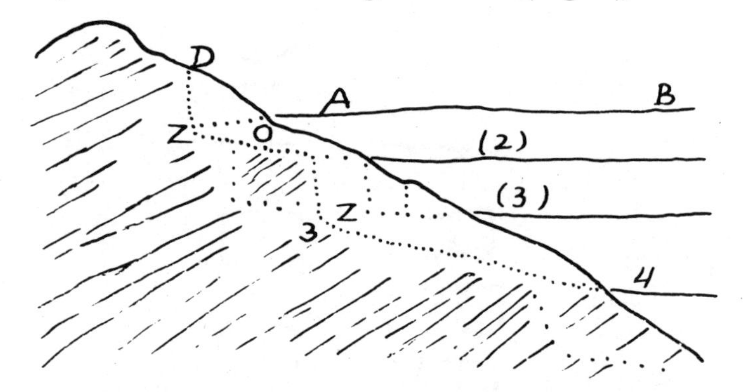

|

39e The action of sea A.B. will be to eat in the land in line of highest tidal action. this will at length be checked by increased vertical ⟨height⟩ thickness (DZ) of mass to be removed & from the resistance offered to the greater lateral extension of the waves. by the part beneath the band of greatest action not having been worn away. — If the level of the sea was to sink by very slow & gradual movements to line (2). The part (O) which was before beneath band. of greatest action. would now by degrees be exposed

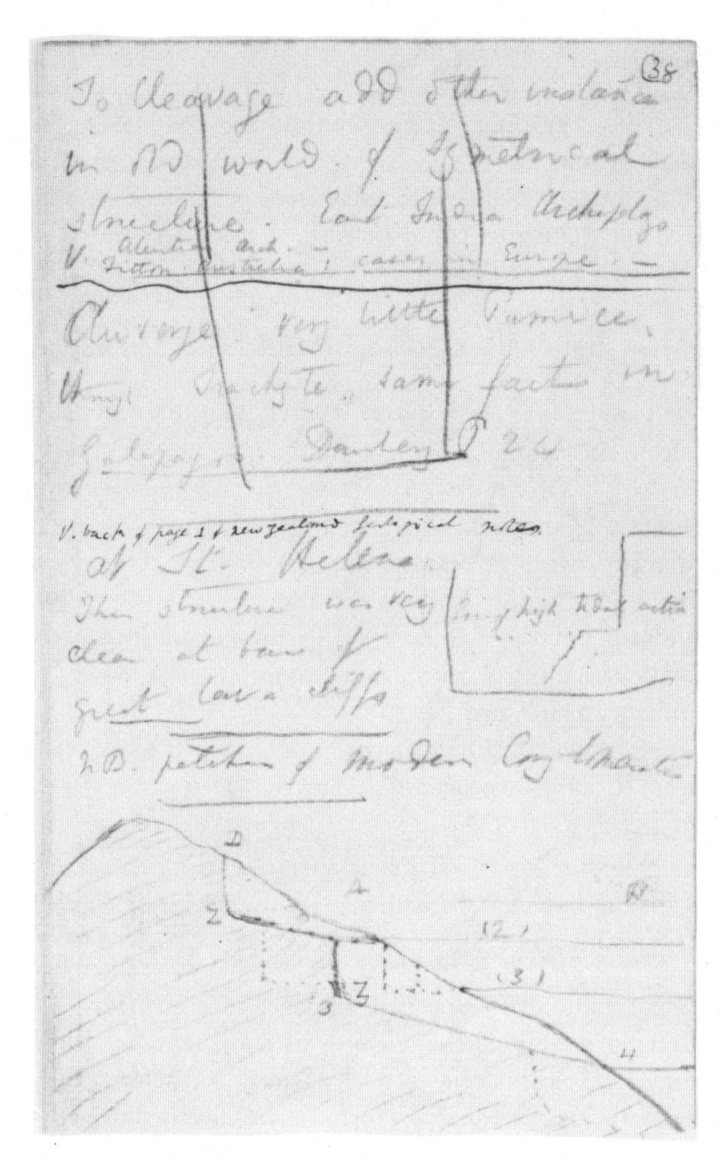

Page 38e with sketches of a possible explanation for the appearance of a section of lava cliffs along the coast at St Helena.

to it, & the result would [be] a uniform slope to base of cliff (Z). to which point the waves would not reach. If now the ocean should suddenly |

40e fall, (3) the case would be as at first. & according to the greater or less time of rest. so would the size of the triangular mass removed vary. — The gradual rising continuing. a another [*sic*] sloping platform would be made, & so on. — This is grounded on the belief of constant rising with successive periods of greater activity & rest. — Such changes could be shown (as represented), along line of coast. — [Fig. 2] Mem San Lorenzo; Valley of Copiapò & parts of coast of Chile. —

Must first explain [top of] tidal band of action. |

41 This case differs. I think. from Patagonian steps, because the deposition & accumulation is brought into play

As in Ocean & Air; there are [likewise] differences of temperature [at equal distances from centre of rotation] & a ⟨circulation owing⟩ rotation in fluid matter of globe. must there not be a circulation [however slow & weak.]; [(cause of not accumulation of Coral limestone in intertropical)] hence varieties of substances ejected from same point. & changes. [(changes in variation?)] as in Cordillera. —

From poles to Equator current downwards & to West. — From Equator to poles. nearer the surface & to the Eastward. — If matter proceeds from great depth. from axis to surface must gain a Westerly current: — If great changes of climate have happened. hurricane in bowels of earth cause: — does not explain cleavage lines./ possibly general symetry of world. — |

42 I feel no doubt. respecting the brecciated white stone of Chiloe, after having examined the changes of pumice at Ascension

In Calc: sandstone at Ascension, each particles coated by pellucid envelope of Lime. — form resembles the husks at Coquimbo: in that case, may not central and rather differently constituted lime have been removed? — As shell out of its cast which, although not very intelligible is a familiar case: If refiltered with other matter how very curious a structure: Have shells ever casts alone in Calc[areous?]. rocks!? — if so case precisely analogous: fragments instead |

43e Peak of Teneriffe. also Cotopaxi has a cylinder placed on the rim of conical crater: at Teneriffe Wall of Porph. Lava with base of Pitchstone; Mem Galapagos. chiefly red glassy scoriæ. — could walk round base: — not universal: could not climb up many parts, in James Is^{d}. — Mem St Helena — All Trachytic. — ⌈Daubeny⌉[51] P. 171. Vol I. Humboldt[52]

There is long discussion on Pumice ⌈& Obsidian:⌉ in the I Vol. Humb:[53]

There is rather good abstract of Humboldt. S. American Geolog. in Daubeny. P. 349[54]

Admirable little table showing long periods of great violence volcanic. from Humboldt: Comparison P 361. Daubeny[55] |

44e Von Buch is very strong about Trachyte being the most inferior rocks[56] — The stream at Portillo Pass example of do? 〈[Poor?]〉

Daubeny good account of ejected granitic fragments. P. 386[57]

⌈Mem. Lyell's fact about sulphuric vapours in East Indian Volcanos⌉[58]

Gypsum

Andes |

45e Mem. Beechey. account of regular change in soundings. on approaching the coast of NW. America P. 209–13 P & 444 ⌈(Yanky Edit)⌉[59]

〈I think〉 At[60] Ascension, the laminæ changes in rocks. connected with & alternating with obsidian must clearly be chemical differences. & not those of rapid cooling &c &c

My results go to believe that much of all old strata of England. formed near surface: Mem Patagonian pebbles beds, most unfavourable to preservation of bones &c &c — Yet 〈silicified〉 turn over[61] |

46e Silicified wood. Cordilleras, Chiloe. &c seems the organic structure most easily preserved. —

Mr Conybeare introduct to Geolog — "Between the height of same beds, deposited in different basins; little or no relation appears to ⟨exist⟩ be made out, but in those belonging to the same district there seems. I think, little ground for skepticism, as to the general truth of the proposition." —[62] If such can happen in troubled England; the more minute equalities |

47 of elevation, may well be preserved at Patagonia. The English fact is astonishing consult book itself. P. xx: same fact is indeed shewn [?][63] by the parallel bands of formations on any Geolog Map: Quoted from Daubeny P 402:[64] likewise, mean height of tertiary. being less than secondary: — consider arguments for oscillation of level independent of mineralogical nature & dependent: & then how wonderful level [of same beds] should have been kept; it shows that throughout all England, whole surface oscillated equably. — |

48 These facts become easy if we look at the action as a deep & extensive movement of viscid nucleus, which in any one country would produce equable effects. — [though so immense to short breathed traveller] Mountains, which in size are grains of sand, in this view sink into their proper insignificance; as fractures, consequent on grand rise, & angular displacement, consequent of injection of fluid rock. —

Try on globe. with slip paper a gradually curved enlargement |

49 see its increased length. which will represent the dilatation, which dilated cracks must be filled up by dikes & mountain chains. —

Introduce part of the above in Patagonian paper; & part in grand discussion

Consult. reconsult Geolog. Map of Europe |

50 Consult charts for distribution of pebbles. — Plains. off coast of Patagonia. — British channel &c &c.

There is a Hill. near Copiapò which is asserted to make a noise, — My impression. is not very distinct, from some of the lower orders; it was connected with movement of sand. — it is called "Bramidor" (?). — it was a strange story; I believe it was necessary to ascend the hill, — but my recollection is imperfect & was recalled by note in |

51 Daubeny. P. 438., of similar fact near the Red Sea. — which occurred in a sandy place. — (the sound was long & prolonged).[65] NB, Is it generally known. the acute chirping sound produced in walking over the sand: I am nearly sure, it is necessary to ascend the hill. —

The absence of Second form, except near submarine Volc: in harmony with the prevailing movement being one of elevation alone. — In England much subsidence: hence difference; action on land different |

52 Volney, P 351. Vol I. woody bushes, ⌈gazelles⌉ hares, grasshoppers & Rats. characteristic of the deserts of Syria ⟨chara⟩ ditto for Patagonia, especially rocky parts of central Patagonia[66]

Does Andes in Chili. separate geographical ranges of plants. V. Lyell. Chap XI Vol II.[67]

Urge the entire absence of any rock situated beneath low water in the Southern ocean not being buoyed with Kelp. — |

53–54e ⌈not located; entry on stub of page 53 reads, "With respect to degrad"⌉

55e Strong currents off the Galapagos. — strata must be accumulating which like the secondary strata of England, ⌈besides ordinary marine remains⌉ may contains ⟨shells few corals Tortoise⟩ ⌈remains of Amphibia, exclusively.⌉ & Turtle bones. & the bones of ⟨two graniniverous⟩ a herbivorous lizard. — from[68] the action of torrents. ⌈marine⌉ Tortoise & other species of large lizard. — There would probably be no other organic remains. — |

56e On Pampas looked in vain for a pebble of any sort; not one was found. — Miers saw them near?[69]

Mem. La Condamaine on the Amazons.[70] Consult

Insist on the frequency of dikes in Granitic countries, enumerate cases. — M. Video exception, but even there, hills of Basalt & other Volcanic rocks. Bahia, Rio de Jan:. B. Oriental? level surface not disturbed. — Whole West coast. Chonos to Copiapo. — Sydney. K.G. Sound. C. of Good Hope. — ⌈Carnatic |

57 It has been common practice of geologist.⌋[71]

Lyell considers (P 84 Vol III.) whole of Etna series of coatings;[72] hence it will be necessary to state all arguments for believing that there must be a central core of melted rock — I think the strongest is the consideration of the state at a grand eruption when whole summit of mountain is blown off; & again when in great crater. different little craters are all burning, surely there must be ⌊somewhere⌋ below a field of fluid rock. — In the discussion it will be better not to refer to Lyell. but merely to |

58 state these reasons, & saying that they refer to central nucleus & that envelopes no doubt existed. These higher portions probably formed Isl^ds from which proceeded pebbles & on which trees grew. — ⌊(?)⌋[73] Are not the dikes in upper strata quite different from the Porphyries: certainly appearance leads me to believe mere fissures filled up. — the appearance will here be the strongest argument: — ʢ Consider causes for subaqueous crater being of diff: form subaerial one? — In former not so much; or no rapilli;[74] & from action of water probably not so much aluminated. |

59 As argument in favor of lines of anticlinal violence crossing lines of crater, ⟨arg⟩ state that all the great Volcanos. have been elevated considerably. which shows an afflux of inferior melted rocks to those parts.

Are not the dikes generally vertical? if so posterior to elevations? & not sources of lava streams. — ⌊Urge not tilted strata. —⌋[75] It will be well to urge the case of St Helena, where dikes certainly have not been points of eruption.

Nobody supposes that all the dikes in Cornwall or in the coal measures have been conduits to volcanoes. — |

60 Talking of the cricket valley ⌊the most remarkable feature in the structure of Ascension⌋[76] give as an example the great subsidence at the famous eruption of Rialeja, & the more true analogy from the Galapagos —

Mr Lyell. P. 111 & 113. ⌊seems to⌋ considers that successive terraces mark as many distinct elevations;[77] hence it would appear he has not fully considered the subject. —

S. America in the form of the land decidedly |

61 bears the stamp of recent elevation. which is different from what Mr Lyell supposes.[78]

Lyell P 116 Vol III, says that in N. Pliocene formation of Limestone, casts of shells, as in some older formations:[79] Mem the envelopes at Coquimbo. the analogy is now perfect

⟨The grand propulsion of fluid rock. which elevates a continent⟩

We are more abound to take analogy of movements of W coast in explaining plains because such are found in perfection on that side. — |

62 Add from M. Lesson. character of Flora to New Zealand, which agrees with St Helena in being unique, yet no quadrupeds. —[80]

Is the white matter beneath pebbles. the degraded matter of such pebbles extending to seaward, the alternating with such matter at St Julians looks like such? — destructive to animal life. — Patagonia. |

63 In the Chonos Islds we must imagine bituminous shales have been metamorphised, as in Brazil feruginous sandy ones have undergone the same process. —

Neither lakes or Avalanches (Glaciers very rare) to cause floods in valleys, which must aid in preserving the terraces ⌈Molina's Case⌉[81]

At Vesuvius. Vol III P. 124. Lyell. dikes have a parting of pitchstone; which is described as very rare.[82] Mem. St Helena; probably more abundant in this case from intersecting a mass probably cold & not warm as sides of a crater as Vesuvius. — |

64 There may have been oscillations in the upheaval of Andes. — but as long as all below water no evidence — The depth of shells (which being packed. in beds) lived there, makes it very doubtful whether they could have lived in so deep a sea. — Perhaps agrees with formation of pebbles & vertical trees

Grand Seco at B. Ayres; mention about the deer approaching the wells. — the effect of Salt water of the Salado. — Mem. in Owens Africa it is mentioned that the Elephant came |

65e [line cut out at top of page] towns driving by the want of water. —[83] I believe in all flat countries. years of drought are common. — Mr Lyell has mentioned the drifting of carcases putrid.[84]

In Rio paper. when discussing probable rise of land: Mention M. Gay's fact about shells:[85] Hibernation of fresh water Shells. multitudes. —

The question of shell's concretions, living only in that spot & being cause of concretion, or being only preserved in that part. having lived over whole bottom is important; because in this latter case we cannot judge whether such fossils lived in groups or not. |

66e Ferruginous veins of this figure (A) in sandstone: evidently depend on a concretionary contraction: the fact is in alliance with those balls at Chiloe full of sand. — the ⟨scale⟩ ⌈quantity of iron⌉ being there in excess. — If veins (A) are secretionary, so are all those plates in Australia. New Red Sandstone. at Bahia in modern sandstone. a circle,  , had in its middle a short ⌈fissure⌉ vein terminated each way, which little vein was like the rest of these thin veins which project outwards. — |

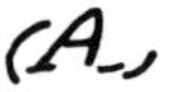

67e In Patagonia. the blending of pebbles & the appearance of travelling may be owing to successive transportal from prevailing swell, (as Shingle travels on the Chesil bank. V. De la Beche).[86] Ask Capt. F.: R:[87] how the swell, generally & during gales would tend to travel on a central line of Patagonia. ⌈NB. Mr Lyell P. 211 Vol III. talks of line of cliff marking a pause⌉[88]

When mentioning pumice of Bahia Blanca, mention black scoriaceous rocks of R Chupat. & fall of Ashes of Falkner, ? how far is the distance?—[89] |

68e Fossil bones black as if from peat. — yet cetaceous bones so likewise ⌈of miocene period⌉. — Mem Bahia blanca P. 204 Vol III. Lyell[90]

Owing to ⌈open⌉ faults in mountains: to elevated strata in eocene lakes of France, & unequal action of Earthquakes ⌈on Chili & delta of Indus⌉. my belief in submarine tilting alone, must be modified. ⌈Moreover, the Volcanos from sea there burst out, after rise from sea: ⟨As did⟩ as did those aerial Volcanos in Germany⌉

In the Valle del Yeso it is probable that point of Porphyry has been upheaved in a dry form

It is clear the forces have acted with far more regularity |

69 in S. America: in France we have freshwater lakes unequally elevated, which movements if present in the Andes, would have destroyed regularity of slope of valleys. — All my observations of period [& manner] of elevation Volcanic action, must be more exclusively confined to that country

Read description of channels or grooves in rocks at Costorphine hills. to compare with Galapagos. — Chiloe. M. Hermoso. & Coral reefs (imperfect in latter). |

70 Lyell. Vol I. P. 316. Earthquake of 1812 affected valley of Missisippi & New Madrid & Caraccas. —[91] Is this mentioned by Humboldt in his account of extensive areas. —[92]

P. 322 In any archipelago. & neighbouring Volcanos. eruption from [more than] one orifice does not occur at same time: this is contrasted to contemporaneous action over larger spaces of the globes & "periods" of increased activity. —[93] such as that of 1835. —

State the three [or 4] fields of Earthquakes in Chili: — |

71 Chiloe. Concepcion. Valparaiso (Copiapò & Guasco). yet whole territory vibrates from any one shock —

In S. America — continuity of space in formations & durability of similar causes go together. add. ⟨"from⟩ "in the same line" to "from the epoch of Ammonite to the present day.

at Mauritius. (consult Bory[94] [dip of strata on East]) cannot believe in a great explosion, nor would sea remove more internally than externally — I did not see any number of dikes in the |

72 cliffs. — wide valleys. — central peak small; yet great body of lavas have flowed from centre —

Pisolitic balls occur in the Ashes which fill up theatre of Pompeei (?). — Such have been seen to form in atmosphere. — Mem. Ascencion. concretions & Galapagos. —

Humboldts. fragmens.[95]

Read geology of N. America. India. — remembering S. Africa. Australia. . Oceanic Isles. Geology of whole world will turn out simple. — |

73 Fortunate for this science. that Europe was its birth place. — Some general reflections might be introduced on great size of ocean; especially Pacifick: insignificant islets — general movements of the earth; — Scarcity of Organic remains. — Unequal distribution of Volcanic action, Australia S. Africa — on one side. S. America on the other: The extreme frequency of soft materials being consolidated; one inclines to belief all strata of Europe formed near coast. Humboldts quotation of instability of ground at present. day. — applied by me geologically to vertical movements.[96] |

74 In Cord: after seeing small Bombs. without a vesicle. we may consider appearances of eruption at bottom. — solution under high pressure of gazes. especially the most abundant. Sulp. Hyd: Carb: A. Mur: A. = (& this effect of water thus holding matter in solution must be great: & in the fact of bombs in tufa there is proof of such gaz) steam condensed. — Perhaps these mighty changes might go on. & not a bubbles on the surface bespeak the changes. —

[metallic veins solution of silex & many other phenomena.][97] |

75 I do not believe that the extraordinary fissures of the ground at Calabria were present at the Concepcion earthquake. — expatiate on difficulty of evidence about eruptions of Volcanos. (where there are no country newspapers) — At the Calabrian earthquake things pitched off the ground. Ulloa states that Volcanos!! were in eruption at time of great Lima earthquake[98]

In the Chili earthquakes if rise was more ⟨than⟩ inland than on coast it would be invariably discovered; this may be mentioned with general slope of the country; (perhaps generally over whole world) |

76 Yet eruptions ⟨both⟩ at sea (as wells as in the Cordillera), they may be considered as accidents (if part of a regular system can be called accidental; the proportional force of crust of globe & injecting matter on the great rise). —

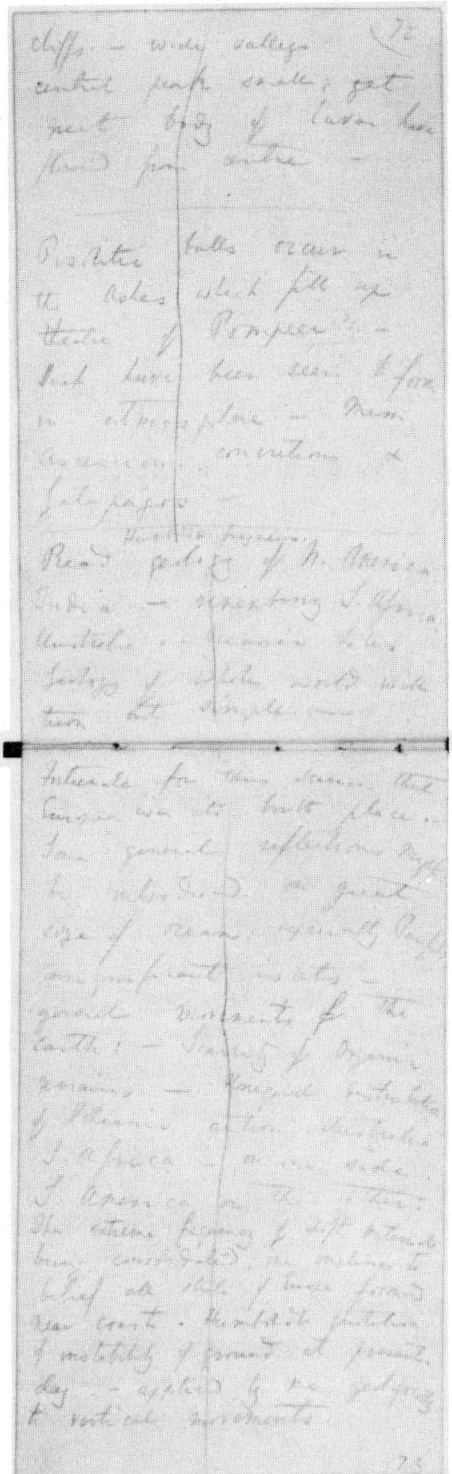

Pages 72–73 on geological topics, with a reference to Alexander von Humboldt's *Fragmens de géologie et de climatologie asiatiques* (1831), and the assertion that the 'Geology of whole world will turn out simple.—'

The great rains which attend severe Earthquakes [1822 ? 1835?] alone, (& the general belief in N. Chili, where rains are so infrequent; so as to exclaim [as I have heard] how lucky! when they hear of a place having a pretty severe shock). are much more curious |

77 & perplexing. than those that attend Eruptions: Mr P. Scopes explanation of low Barometer?[99]

In a subsiding area. we may believe the fluid matter instead of afflux (always slightly oscillating as that of a spring) moves away.— Will geology ever succeed in showing a direct relation of a part of globe rising, when another falls. — When discussing connection of Pacifick & S. America. — |

78 Volcanos must be considered as chemical retorts. — neglecting the first production of Trachyte. look at Sulphur. salt. lime, are spread over [whole] surface; how comes it they do not flow out together? How are they eliminated. — [Sulphur last. —] Metallic veins likewise must separate ingredients if we look to a constant revolution. — Are we to consider that the dikes which so commonly (state facts) traverse granites, are granitic materials simply altered by circumstances; & not in chemical nature, or has a subterranean fluid mass itself changed. — No. — |

79 Yet the fluid granitic mass under less pressure might have its [proportional] particles altered. —

With respect to Volcanic theory. I want to ground, that the first phenomem. is an inward afflux of melted matter. — Volcanos perhaps may be admittance of water, through the rent strata: [Mr Lyell considers that Plutonic rocks are generated as often as Volcanic. I consider latter as accidental on the afflux of the former. —][100]

Ascension. Vegetation? Rats & Mices. At St Helena there is a native mouse. |

80 Did wave first retreat at Juan Fernandez: the first great movement was one of rise (any smaller prior ones might have been owing to absolute movement of ground). Michell (Philos:

Transacts) [seems to] consider that fall first movement (as in Peru 1746). —[101] At great Lisbon Earthquake Loch Lomond water oscillated between 2 & 3 ft. (as in Chili lake). Therefore motion of sea ought to be considered as a plain movement communicated to it as well as by the vertical as lateral movement. — At first one would think movement. owing to water keeping its level whilst land rose up & down. — But from above reasons, do not think so |

81 also elevating Earthquake of Valparaiso. (1822) no great wave on record. — [also neighbouring sea must partake in absolute movement] Moreover wave [with same general character] reaches far beyond coast, which has been raised. — It must be considered as an oscillation, from violence. Is it not same as swell travelling across Pacifick. — excepting in number of waves & in wind, instead of sea's bottom being in motion what difference? In watching heavy swell, sea retreats & then breaks: i e to form a wave in ocean. is not this [Fig. 3]  c a a a c form present, i e a part below [mean] level before the higher part. —
Does the |

82 sea fall on banks as a Bore wave rushes up? (NB. Earthquake wave is an oscillation, body of water manifestly does not travel up. —) If these view are right the coincidental retreat at Portugal & Madeira (Lyell. vol I. P. 471) is explained.[102] also the similar fact at Concepcion? Read the various accounts & see if fall is not the first very evident movement. — The swelling first on beach I cannot understand, without (c^s raised above a^s). — |

83 In great Calabrian wave did not sea break first? I can imagine from local form of coast (as seen in swell) the undertow & overfall must vary proportionally

Partial shrinking after elevation in perfect conformity with ⟨Mr Lyell's⟩ idea of an injected mass of fluid rock[103]

In Patagonia plains. long periods of rest & vice versâ more likely to be coincidental than single elevations along whole line of coast |

84 Darby mentions beds of marine shells on banks of Red River Louisiana. V. Lyell. Vol I. P. 191[104]

State at St Helena. pebbles entirely coated with Tosca. which implies motion in the [loose] bed of pebbles. (On a sea beach under a cascade, one can understand pebbles thus coated. — The motion is most wonderful, from chemical attraction, as a blade of grass penetrating by action of Organic power a lump of hard clay — |

85 In the History of S America we cannot dive into the causes of the losses of the [species of] Mastodons. which ranged from Equatorial plains to S. Patagonia. To the Megatherium. — To the Horse. = One might fancy that it was so arranged from the forsight of the works of man

Feeling surprise at Mastodon inhabiting plains of Patagonia is removed by reflecting on the nature of the country in which the Rhinoceros lives in S. Africa: the same caution is applicable to the Siberia case |

86 We must not think alluvial plains [always] most favourable; In what part of the globe are there such vast numbers of wild animals. both species & individuals as in the half desert country of S. Africa. It would be well to quote Burchell. V. where the Rhinoceros was killed. — [105]

In Patagonia, are all beds same age? is white substance triturated Porphyritic rocks (mem white tufas with purple Claystones of P. Desire). = Where talking of such substances being worn into channels. |

87e mention submarine channels. such as that in front of St^{s}. of Magellan

In Chiloe curvilinear strata subsidence. — The sudden increased dip is not parallel case to Isle of White. but rather to one out of a series of faults. [Fig. 4]

In Cordill: should basal lavas be called Volcanic or Plutonic The cellular state of all the Porphyry specimens, must be well examined

At M. Video [facts of Passages marked by do.] discuss quartz veins, there contemp — yet similar ones in Clay. Slates contemporaneous others subsequent. as in dikes. |

88e In Granite great crystals arranged on sides. V. Lyell P 355 Vol III. constitution of veins, is there said granite in close contact varies in nature,[106] — Does not granite at C. Tres Montes become more siliceous in close contact? — [Cordillera???] Porphyry at Valparaiso; Epidote —

Must we look at regular greenstone cones at S. T. del Fuego as nucleus of a Volcano or as an injected mass. — From conical form I incline to ⟨latter⟩ former; & thus occurring in groups. — As these greenstone rocks are seen to graduate into granites |

89 the ⟨conta⟩ passage from lava to Granite is much more perfect. than in believing mere agency of dikes: & indeed when do these dikes lead to a conical mass. will this conical mass be granite? Why not more probably greenstone? What probable origin can be given to the numerous hills of greenstone? —

Daubeny. P 95. Glassy & Stony Pearlstones alternate together in contorted layers:[107] Mem: Phillips Mineralogy some such fact stated to exist in Peru.[108] — Ascension |

90 At Ischia there is a pumiceous conglomerate with small & large fragments, nature of which is doubtful. P. 180.[109] I think my Ascension case very doubtful. —

In Iceland Bladders of Lava are described, & many minute craters as at Galapagos. Sir George Mackenzie must be worth reading[110]

Some earthquakes of Sumatra no connection with a neighbouring Volcano of Priamang. — Marsden Sumatra.[111]

M. De Jonnes seems to |

91 think that Volcanic eruptions form foundations for Coral reefs. —[112] does he mean in contradistinction to sand??

B. Roussin states that generally in North part of Brazil. ⟨gravel becomes⟩ sand less & gravel more common. the shoaler the water & nearer the Banks[113]

Is there not a sudden deepening on E. coast of Africa. as at Brazil |

92 [blank]

93e What is nature of strip of Mountain Limestone in N. Wales. was it reef. — I remember many Corals?? Breccia — Stratification?

Anomalous action of ocean. — at Ascension. (where occasionally most tremendous surf & loose sandy beach) deposits ⌈calcareous⌉ encrustations; At Bahia ferruginous. — At Pernambuco (great swell & turbid water) organic bodies protect like peat reef of sandstone. — Corals, & Corallina survive, in the most violent surfs: in both latter cases become petrified, & increase. — In Southern regions every rock is buoyed by Kelp, now Kelp sends forth branching |

94e roots which must protect surface: On ⌈hard⌉ exposed rocks near Bahia, whole surface to where highest spray (there pale green confervæ) coated with living beings; In smooth seas (& even turbulent as at St Helena) I have mentioned point of greatest action; I now having seen Pernambuco believe much is owing to protection of Organic productions. = Yet everywhere on coast (Il Defonsos ⌈Kelp⌉) rocks show signs of degradation; (soft substances worn into bare cliffs evident); the action is anomalous; It is wonderful to see Coral reef — or confervæ in the breakers or in waterfall: Excepting by removal of large fragments by mere force of waves: & action on upper tidal band, I do not |

95e see how to account for oceans power. — excepting when pebbles are brought into play; most manifest example of degradation I ever saw on beach near Callao. — From Sir. H Davy experiment on the copper bottom. we see a trifling circumstance determines whether an animal will adhere to a certain part.[114] Apropos to question does animal adhere to rock because it does not decompose. or vice versâ. Clay slates unfavourable to attachment of many bodies |

96e ⌈blank⌉

97e Beechey. — changes in bottom in NW coast of America. from shingle to sand &c &c. ⟨Vol II⟩ P. 209. 211. 213. 444 ⌈Yanky edition⌉[115]

Shores of Pacifick, as compared to whole E. America. ⟨East⟩ Africa. Australia. profoundly deep: a great fault or rather many faults. —

Necessary form; as long as coast line fixed. — [Fig. 5][116]

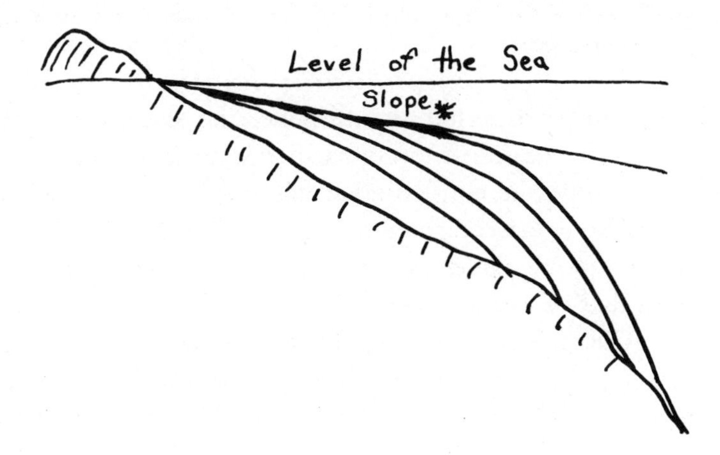

*Slope necessary for seaward transportal of drift matter. — |

98e Give various cases. [Fig. 6]

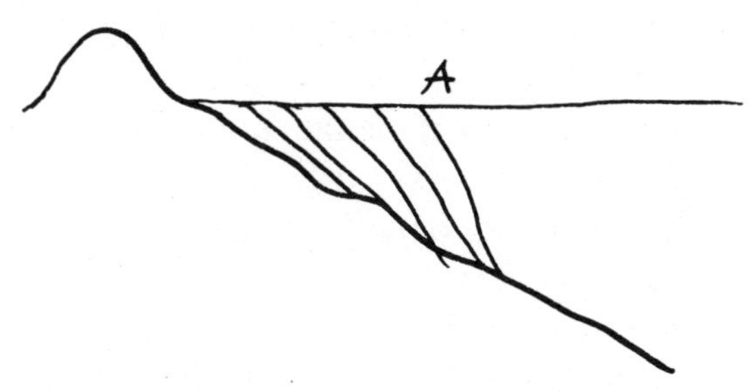

A advancing coast to Seaward.

Retreating case in excess as first case.

When discussing Falkland soundings introduce this discussion.— Brazil bank: (& I believe SE coast of Madagascar. where a 40 line runs at equal distance?) 1st cases. —[117] |

99e The terraces in Valleys of Chili may be with much truth compared to the step = formed streams of lava at St Jago. C. de Verds

Quartz pebbles in the Cordilleras look as if some peaks elevated.—

Greywacke as a general fact absent in T. del Fuego, excepting in Port Famine

Mr Sorrell says that numerous icebergs are commonly stranded on shores of Georgia [Lat° ( )], he has rocks on surface applicable to Patagonia.[118] |

100e During a period of subsidence the shinglle of Patagonia would become more or less interstratified with sediment. — [& escarpment worn away like english escarpment][119]

The great conglomerate of the Amazons & Orinoco mentioned by Humboldt under name of Rothe-todte-liegende is perhaps same with that of Pernambuco?[120]

Quote Miers about shells at Quillota[121]

Lyell, states that contact of Granite & sedimentary rocks, in Alps becomes metalliferous. Vol III Latter Part[122] |

101 Are there Earthquakes in the Radack & Ralix Isl^ds?

In my Cleavage paper Dr Fittons Australia case must be quoted at length.[123]

The Lines of Mountain appear to me to be effect of expansions acting at great depths (mem: profound earthquakes), which would cause parallel lines, but the rectangular intersections are singular —

M. Lesson considers the Sandstone & Granite districts to be separated by profound valley[.] Sydney. —[124] |

102 Lesson Zoologie[125]

Grand tertiary formation of Payta: N. part of New Zeeland entirely volcanic!! New Zeeland rich in particular genera of plants: All St. Catherine & coast Granite: P. 199; Falkland account of cleavage differs wonderfully from mine: phyllade covered by quartzose sandstones: refers to broken hill described by Pernetty: account of streams of stones agrees with mine. — At Conception, cleavage E & W! at Payta. talcose slates, do at latter place. sandy. sandstone with gypsum, covered by limestone with recent shells 200 ft, how exact agreement with Coquimbo; |

103–104e [not located]

105e Is^{d} near coast of America not reached. Juan. Galapagos. Cocos — Ulloas voyage

North of Callao, the country, to the distance of 3 or 4 leagues ⌈from the coast⌉ may be concluded to have been covered by the sea — judge from the pebbles such as those on the beach — "This is particularly observable in a bay about five leagues North of Callao, called Marques, where in all appearances not many years since, the sea covered above half a league of what is now Terra Firma & the extent of a league & a half a long the coast. The rocks in the most inland part of this bay are perforated & smoothed like those washed by the waves, a |

106e sufficient proof, that the sea formed these large cavities," &c &c &c Vol II. Chapt VIII. p. 97[126]

at Potosi the veins run from North ⟨inclining⟩ to South. inclining a little to the West: the veins which follow this direction are thought by the ⟨oldest⟩ most intelligent miners to be the richest Vol II 147[127]

Shells at Concepcion 50 toises above the sea. = talks of them being packed clean. & without earth. — Moreover that such do not occur on the beaches. Perhaps these facts attest a ⟨more⟩ decided elevation of sea's bottom. beds of shells. 2–3 toises thick.— Vol II. p. 252[128] |

107 Urge cliff form of land, in St Helena. Ascension. Azores. (⌈sandstone first gives⌉ half demolished craters). — worn into mud & dust. — connection with age, & agreement with number of craters. No cliffs at Ascension (or modern streams of St Jgo) yet no historical records of eruptions how immense the time!! How well agrees with number of Craters! — At S. Cruz. there is no occasion to wonder what has become of the Basalt. Gone into fine sediment Look at St Helena!! — |

108 There are some arguments which strike the mind with force. — the exact yearly rise of the great rivers prove better than any meterological table the precise periods over immense areas. (& the counterbalancing variations) of rain. = The Bulk of sediment ⌈daily⌉ yearly brought down by every torrent proves the decay atmospheric

of the most solid rocks. — The grand cliffs of a thousand feet in height, of the solid lavas. — proportionally high to age. (we do not wonder to see tertiary plains consumed) Where slope [plainly] indicates former boundary. (as in other unworn |

109 islands) we take in at once the stupendous mass which has been corroded. — If man could raise such a bulwark to the ocean, who would ever suppose that its age was limited? Who could suppose such trifling means could efface & obliterate so grand a work? — In valleys one is not sure whether fissures may not have helped it, or diluvial waves. but when we see an entire island so encircled, the one slow cause is apparent. [I confess I never see such islands whose inclination natural [out limits?] [illeg.] deepest astonishment.] Perhaps scarcely a pebble might remain to [tell?] of these losses. — |

110 Cause of chimney. to crater. as at Galapagos. St. Helena. —

[Fig. 7]  effect of heat on inner wall, hence resists degradation longer than outer parts. —

The common occurrence of a breccia of primitive rocks between that formation and the secondary (stated in Playfair to be the case p. 51).[129] presupposes an elevated country of granite, not greater for all Europe, than from the Plata to Caraccas, which is all of granite: |

111 In discussing circulation of fluid nucleus, — the similarity of Volcanic products [over whole world] argument, as well as separating causes by water. — Or rather begin & explain how water separates. — (intertropics at present fix lime). ⟨Also Volcanos separate.⟩ Volcanos blend all substances together; & products being similar over whole world. general circulation. But Volcanic action separates some sulphur (perhaps lime) salt. & metallic ores. — which mingling & separating is well adapted to |

112 use of mankind. — ⟨Hutton show⟩[130] Earthquakes part of necessary process of terrestrial renovation & so is Volcano a useful chemical instrument. — Yet neglecting these final causes. — What more

awful scourges to mankind than the Volcano & Earthquake. — Earthquakes act as ploughs Volcanos as Marl-pits: |

113e Consider well age of Bones. = slowness of elevation proved at St Julian. = do not these bones differ as much nearly as the Eocene. = Should Mr Owen consider bones washed about much at Coll. of Surgeon's?[131] I really should think probably that B. Blanca & M. Hermoso contemp:. — Inculcate well that Horse at least has not perished because too cold: — With discussion of camel urge S. Africa productions. — |

114e I think in Patagonia white beds having proceeded from gravel proved. — curious similarity of rocks of very diff. ages. at Port Desire on plain & interstratified. —

Urge fact of Boulders not in lower strata. only in upper. in accordance in Europe with ice theory. —

Capt Ross found in Possession Bay in 73° 39 N. living worms in the mud which he drew up from 1,000 f[athoms], & the temp of which was below freezing point!!![132] |

115 Remember idea of frozen bottom [or beach] of sea to explain preserved animals. — Mem: stream of water in the country. —

Sir J. Herschel says precip. of Sulph. B. all the infinitesimal cryst. arrange themselves in planes. [Mem silky lustre][133] ask Erasmus. whether electricity would affect this. —[134]

State the circumstances of appearance at Concepcion[.] no sign of elevation. Effects of great waves to obliterate all land marks. — At [the?] first it |

116 would though be easy to see on beach successive lines of sea weed —

Histoire Naturelle des Indes

Acosta. p. 125 of French [?] Edition states that the same earthquake has run from Chili to Quito a distance of more than 500 leagues. A little time after a bad earthquake in Chili; Arequipa in 82 was overthrown, & 86. Lima. next year Quito. considers these earthquakes travel in order. —[135] |

117 If we look at Elevations as constantly going on we shall see a cause for Volcanos part of same phenomena lasting so long. —

The great movements (not mere patches as in Italy proved by Coral hypoth. agree with great continents). |

118 Voyage aux terres Australs Vol. I. p. 54. M. Bailly says. "en effet toutes les montagnes de cette île se developpent autour d'elle comme une ceinture d'immenses remparts; toutes affectent une pente plus ou moins inclinée vers le rivage de la mer, tandis, au contraire, que vers le centre de' l île, elles presentent une coupe abrupte et souvent tailée a pic. Toutes ces montagnes |

119 sont formées de couches paralleles et inclinées du centre d'lile, vers la mer; ces couches ont entre elles une correspondance exacte, et lorsquelles se trouvent interrompues par quelque vallées ou par quelque scissures profondes, on les voit se reproduire a des hauteurs communes sur le revers de chacune des montagnes qui forment les vallées ou les scissures. — M. B. thinks these parts incontestably formed the parts of one whole |

120 burning mountain, & that the central part fell in. — Says posterior craters in centre:—[136] Bailly talks of much granite on all East side of Van Diemen Land.[137]

All the Calcareous rocks which harden by themselves cannot be pure. for if so Chalk would |

121 harden. — Climate.!? or small Proportion of Alum: matter. — all pale cream colour. —

The Brecciated structure of all the Pitchstone (which I have seen) is a kind of concretionary structure, for the interlineal spaces are of diff conts: & even in one case contained lime. — All bear close analogy to Obsidian, & all show chemical action as well as effects of cooling |
[misnumbering, no page 122]

123 In Igneous rocks. — which have the cryst of glassy F. fractured. have been melted with little pressure. & perhaps cooled suddenly. —

As the rude symmetry of the globe shows powers have acted from great depths, so changes, acting in those lines. must now proceed from great depths. — important. — |

124 Decemb 10. 1802. Earthquake at Demerara. The earthquakes "seem to arise from some efforts in the land to lift itself higher & to grow upwards; for the land is constantly pushing the sea (which of course must retain same level) to a greater distance". — Afterwards speaks of this phenomenon in connection with "shooting upwards" of the ⟨ground⟩ land in the W Indies. — p. 200. Bollingbroke voyage to the Demerary[138] |

125 Earthquakes at St Helena. 1756. June 1780, Sept 21st. 1817. — p 371. Webster Antarctic veg:[139]

Study Ulloa to see if Indian habitation above regions of vegetation. — ⌈I can find nothing.⌉[140] Mem Carolines quotation from Temple[141]

Urge the mineralogical difference of formations of S. America & Europe. — If great chain of Volc. had been in action during secondary period how diff. would the rocks have been. The red Sandstone of Andes fusible? |

126 no. mad dogs. Azores. although kept in numbers. p. 124. Webster[142]

Consult W. Parish.[143] & Azara.[144] about dry season[.] 1791. seen commonly bad over whole world. ⌈(Was it so in Sydney, consult history? Phillips.⌉[145]

1826.27.28. grt. drought at Sydney. which caused Capt. Sturt expedition. —[146]

§Another one in 1816 (?). — |

127 Mr Owen's curious fact about Crust [word begun: Br...] in Brine.[147] Springs. (Henslow)[148]

Speculate on neutral ground of 2 ostriches; bigger one encroaches on smaller. —[149] change not progressif⟨e⟩: produced at one blow. if one species altered: ⟨altered⟩ Mem: my idea of Volc: islands. elevated. then peculiar plants created. if for such mere points; then any mountain. one is falsely less surprised at new creation for large. — Australias = if for volc. isld. then for any spot of land. = Yet new creation affected by Halo of neighbouring continent: ∤ as if any |

127 Mr Owen's curious fact about Crust[acea] in Brine Springs (Henslow)

Speculate on neutral ground of 2 ostriches; bigger one encroaches on smaller.— change not progressive: produced at one blow. if one species altered: Mem: my idea of Volc. islands. elevated. then peculiar plants created. if for such mere points; then any mountain. one is falsely led supposed at new creation for large.— Australia's = if for Volc. isl[an]d. then for any spot of land. = Yet new creation affected by Halo of neighbouring continent: = as if any

creation, taking place over certain area must have peculiar character: 128

Contrast low limit of Palms, evergreen trees, adjoining arborescent ferns, parasitic plants, Cacti: & with limits of no vegetation at S. Shetland: =

Great contrast of two sides of Cordillera, where climate similar.— I do not know botanically = but picturesque = Both N & S. great contrast. from nature of climate =

Perpetual snow,— subterranean lakes, near Volcanoes, lakes of brine all inhabited;

Go steadily through all the limits of birds & animals in S. America Zorilla:

Pages 127–128 on the species question.

129 wide limits of Waders: Ascension. Keeling: at sea so commonly seen at long distances: generally first arrives:—

New Zealand. gates opening in the history of rats, in the antipodes a parallel case.—

Should urge that extinct Llama owed its death not to change of circumstances; reversed argument. knowing it to be a desert.— Tempted to believe animals created for a definite time:— not extinguished by change of circumstances:

The same kind of relation that common ostrich bears to (Petise. & diff kinds of Fourmillier): extinct Guanaco to recent: in former case position, in latter time. (or changes consequent on lapse) being the relation.— As in first cases distinct species inosculate, so must we believe ancient ones: ∴ not gradual change or degeneration. from circumstances: if one species does change into another it must be per saltum— or species may perish. = This ~~inosculation~~ representation of species important, each its own limit & represented.— Chile Creeper: Furnarius. ~~Caracara~~ Calandria; inosculation alone shows not gradation;— 130

Pages 129–130 on the species question.

128 creation [taking place] over certain area must have peculiar character:

Contrast low limit of Palms, evergreen trees, arborescent grasses, parasitic plants, Cacti: & with limits of no vegetation at S. Shetland. = [150] Great contrast of two sides of Cordillera, where climate similar. — I do not know botanically = but picturesquely = Both N & S. great contrast from nature of climate. =

Perpetual snow. — subterranean lakes, near Volcanoes. lakes of brine all inhabited:

Go steadily through all the limits of birds & animals in S. America. Zorilla:[151] |

129 wide limits of Waders: Ascension. Keeling: at sea so commonly seen. at long distances; generally first arrives: —

New Zealand rats offering in the history of rats, in the antipodes a parallel case. —

Should urge that extinct Llama owed its death not to change of circumstances; reversed argument. knowing it to be a desert. —[152] Tempted to believe animals created for a definite time: — not extinguished by change of circumstances: |

130 The same kind of relation that common ostrich bears to (Petisse.[153] & diff kinds of Fourmillier)[154]: extinct Guanaco[155] to recent: in former case position, in latter time. (or changes consequent on lapse) being the relation. — As in first cases distinct species inosculate, so must we believe ancient ones: [∴] not <u>gradual</u> change or degeneration. from circumstances: if one species does change into another it must be per saltum — or species may perish. = This ⟨inosculation⟩ representation of species important, each its own limit & represented. — Chiloe creeper:[156] Furnarius.[157] ⟨Caracara⟩[158] Calandria;[159] inosculation alone shows not gradation; — |

131 An argument for the Crust[160] of globe being thin, may be drawn. from. Cordillera. rocks. — When beneath water. — together with hypothetical case of Brazil. — |

132 Propagation. whether ordinary. hermaphrodite. or by cutting an animal in two. (gemmiparous. by nature or accident). we see an individual divided either at one moment or through lapse of ages. — Therefore we are not so much surprised at seeing Zoophite producing distinct animals. still partly united. & egg[s?] which become quite separate. — Considering all individuals of all species. as [each] one individual [divided] by different methods, associated life only adds one other method where the division is not perfect. — |

133 Dogs. Cats. Horses. Cattle. Goat. Asses. have all run wild & bred. no doubt with perfect success. — showing non Creation does not bear upon solely adaptation of animals. — extinction in same manner may not depend. — There is no more wonder in extinction of species than of individual. — |

134e Mr Birchell says Elephant lives on very wretched countries thinly covered by vegetation.[161] Rhinoceros quite in deserts. — Much struck with number of animals at Cape of Good Hope

Says at Santos [M Birchel[s?]] at foot of range some miles from shore. rock of oysters quite above reach of tides. — thinks them same as recent species. —[162] |

135e May I not generalize the fact glaciers most abundant in interior channels. there no outer coast. — important effect. — ? Capt. Fitz Roy. — [163]

Limited Volcanic action & limited earthquakes & great but local elevations of the land in Europe — |

136e Urge difference of plutonic rocks & Volcanic metalliferous —

Urge enormous quantity of matter from crevice of Andes — therefore flowed towards it. a mass on each side 3000 ft thick & 150 broad. neglecting Cordillera itself now remaining — |

137e Lyell [⟨p 419⟩ p 428] states that Von Buch has urged that Java volcanos differ from all others in quantity of Sulph. acid emitted:[164] mem: Grand gypseous formation of Cordillera

In describing structure of Cordillera it must be said, that lines of elevation have connected ⟨lines⟩ [points] of eruption[.] give instance of Etna, Stromboli & Vesuvius |

138 Investigate with greater care. vegetation & climate of Tristan D. Acunha. Kerguelen Land. Prince Edwards Isd. Marion & Crozet. L. Auckland. Macqueries. — Sandwich Isd —

Specimens of rocks were brought home in ⌈written over 'by'⌉ Capt. Forster expedition from ⟨Deception Isld.⟩ South Shetland Cape Possession. Syenite ʕ Andite? —[165] |

139 Degrading of inland bays. like St. Julian & Port Desire applicable to Craters of Elevation. — The longer diameter of Deception Isd is six Geographical miles and width 2 & 1/2 miles[166]

S. Shetland. Lat. 62° 55′. ⟨onl⟩ one lichen only production. a body which had long been buried, from rotten state of coffin ⌈buried in a mound⌉ long consigned to the earth. yet body had scarcely undergone any decomposition: countenance so well preserved. that it was thought not to have belonged to an Englishmen. — On 8th of March cove began to freeze. correspond ⌈to September⌉[167] |

140e ʕDid I make any observations on springs at S. Cruz.??? —

Form of land shows subsidence in T. del Fuego, and connection of quadrupeds. — although recent elevation, there may have been great subsidence previously. Mem. pebbles of Porphyry. — Falklands. — off East Coast. — Capt. Cook found soundings. (end of 2d voyage outside coast of T. del Fuego. off. Christmas sound. —[168]

⌈(Think some 60 fathoms, none thicker than thumb⌉ Sea weed said at Kerguelen Isd. to grow on shoals like Fucus giganteus! 24 fathoms deep 24 |

141e under 50. Kerguelen Land, = the way it stands gales = very strong. Stones as bigger than a man's head. — ⌈Kerguelen 40 by 20 leagues. dimensions:⌉[169]

Bynoe informs me that in Obstruction Sound, in the narrow parts which break through the N & South lines the tides form eddies with its extreme force.[170] Yet, no outlet at head. Important in forming transverse valleys

Ice |

142e Sir W. Parish says they have Earthquakes in Cordoba. one of which dried up ⟨all⟩ a lake in neighbourhood of town[171]

Mr Murchison insisted strongly. that taking up a piece of Falkland Sandstone. he could not distinguish from stone Caradoc from lower of third Silurian division — Together with same general character of fossils deception complete. —[172]

Silliman Journal. year 1835 excellent account of N. American geology. Conybeare[173] |

143e Lava in Cordillera & on Eastern plains [by Antuco]. Athenæum April 1836 (p 302)[174]

Coleccion de obras. 2 Vols fol: Buenos Ayres 1836:[175] W. Parish?? [by Pedro de Angelis.][176]

This work is reviewed in present Edinburgh March 1835[177]

Sir W. Parish says. that beds of shells are found on whole coast from P. Indio to Quilmes. & at least seven miles inland.[178] |

144e The Cordoba earthquake a very remarkable phenomenon. showing line of disturbance inside Cordillera: It is not therefore so wonderful that volcanic rocks at M. Video [Volcano in Pampas]

Pasto Earthquake. Happened on January 20th. 1834

Mr Sowerby. younger. says that Falkland fossils decidedly belong to old Silurian system.[179]

Apply degradation of landlocked harbors to Craters of elevation. — |

145e Lyell suggested to me that no metals in Polynesian Islds — .[180] Volcanic plenty in S. America !! Metamorphic |

146 Volcanos only <u>burst</u> out where strata in act of dislocation (NB. dislocation connected with fluidity of rock ∴ [in earliest stage] when covered up beneath ocean). — The first dislocations & eruptions can only happen during first movements, and therefore beneath ocean, for subsequently there is a coating of solidifying igneous rocks which would be too thick to be penetrated by the repeted trifling injections. — Old vents would keep open long after emersion, but improbably so long, that to be surrounded by continent. — change of volcanic focus. — |

147 ⟨it is certain, if strata can be⟩

Problem dislocate strata without ejection of the fluid propelling mass. If one inch can be raised then all can, for fresh layers of igneous rock replace strata. & it is nothing odd to find them injected by veins & masses [Fig. 8]

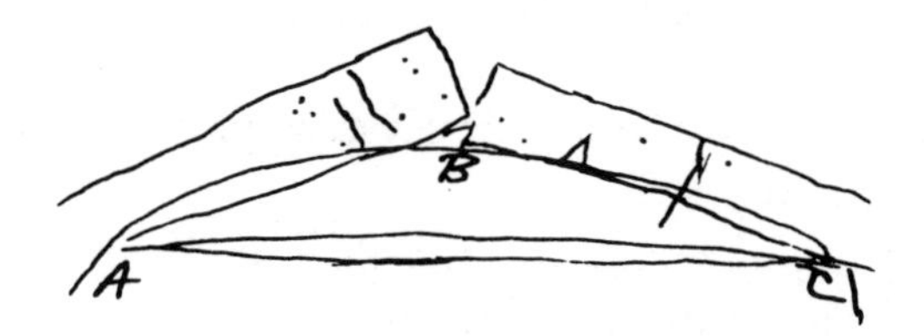

(A.B.C. now grown solid.) |

148 Red Sea near Kosir, land appears elevated. Geograph. Journal p 202 Vol IV[181]

When recollecting Gulf of California. Beagle Channel. — One need never be afraid of speculating on the sea |

149 The 24 ft. elevation at Concepcion. from impossibility of such change having taken place unrecorded must be insensible.

Quantity of matter from Cordillera. horizontal movement of fluid matter not (for instance) expansion of solid matter by Heat |

150 Consider profoundly the sandstone of the Portillo line. — connected with ⟨gneiss⟩. — (Mica Slate) [Fig. 9]

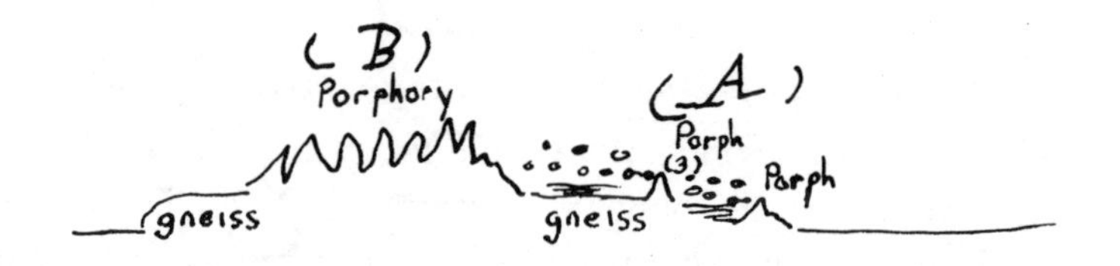

((3) like Bell of Quillota.) (A) in this strata may be older than (B). Most important view Urge curious fact felspar melted gneiss/// Quartz!!! Analogous to Von Buch. Basalt where Basalt. trachyte where trachyte.[182] |

151 There must have been as much conglomerate on West of Peuquenes as on East.

Where gone to.? —

There must have been some conglomerate East of Portillo

Where gone to? Intermediate space protected. —

Oh the vast power of the ocean! |

152 Make a grand analogy between Wealden & Bolivia

Transportal of conglomerate between two ranges mysterious!—

Mem. Subsidence Uspallata of which no trace except by trees |

153 The structure of ice in columns. show that granite when weathering into balls. must exhibit orbicular structure. — When we recollect connection of columnar & orbicular in basalt. —

When we see Avestruz two species. certainly different. not insensible change. —[183] Yet one is urged to look to common parent? why should two of the most closely allied species occur in same country? In botany instances diametrically opposite have been instanced: it is |

154 Let it not be overlooked that except by trees, I could not see trace of Subsidence at Uspallata. —

¿If crust very thick would there be undulation? would it not be mere vibration? but walls & feeling shows undulation ∴ crust thin. — Concepcion earthquake |

155 Draw close Analogy Lake of Cordill: of Copiápò & Desaguadero. — three ridges in Copiapo, as well as in latter. —

According to Mr Brown,[184] a person (whom I met at S.W.P.) the Cordillera extend to near Salta. & not far from Tucama[n]. & at Chuquisaca. half across the continent. — He states plains of Mendoza smooth. Sir W.P. states that in Helm's travels accounts of travelled boulders from the Cordovise range.[185] Signor Rozales tells me at seven oclock Novem ⟨5th⟩ Concepcion most violently shaken by earthquake. but no serious injury. —[186] |

156 ⟨Analysis of Atacama. Iron in Edinburgh. Phisoph. Transactions. = Mem: Olivine. Volcanic product. = ⟩[187]

⟨Did Peruvian Indians use arrows or Araucanians? —⟩

If wood now preserved over world Dicotyledons far preponderant, if so coniferous must formerly have been most abundant tree —

Metamorphic action: ⟨most⟩ coming so near surface most important |

157e There is map of Cordillera by Humboldt in Geolog. Society[188]

Sir Woodbine Parish informs me that town near Tucuman and Salta. towards the Vermejo was utterly overthrown by earthquake with great destruction of human life. — [189] Temple mentions some earthquake at Cordova. — [190] There the Cordova earthquake |

158e in which lake was absorbed. — Earthquakes felt. different case from shore of Pacific. — Isabelle's volcano, many amygdaloids. — [191] Boussingault [(Lyell)] cracks mountains falling in. — [192] Earthquakes at Quito. tranquility [at Mendoza] exception. — [formerly perhaps otherwise] Mendoza never overthrown, — no mountains |

159 Mackenzie has talked of lava flowing up Hill; ⸮what does he mean?)[193] Consult Dr Holland about bubbles. — [194]

No Volcanic action on coast line of Old Greenland, close to W of Jan Meyen Is^{ld}. — Mr Barrow[195] thinks N & S. line connects western isles of Scotland & Iceland. — [Bosh][196] nor on Norway, or Spitzbergen. — Spitzbergen animals (?). |

160 The Hollowness of ⟨sep⟩ Chiloe concretions somewhat analogous to septa. — would particle attracted towards space tend to form ring. [Fig. 10]

motion from within and without

H. Kingdom N. Spain. Vol III p. 113 "Nature exhibited to the Mexicans enormous masses of Iron and Nickel, & these masses which are scattered over the surface of the ground are fibrous. malleable & of so great tenacity, that it is with difficulty that a few fragments can be separated from them with steel instruments."[197] |

161 In R. Brown (Collect: [of F. W.])[198] where the stalactiform masses have layers been accumulated, round knobs, or pushed where soft, or [redissolved?] soft. — /is there any flexure ⟨fr⟩ in the fragmentary jasper. — do undulations (as Hutton says)[199] always come from without. —

[continued from previous page] "True native iron that to which we cannot attribute a meteoric origin & which is constantly found mixed with lead & copper is infinitely rare in all parts of the globe". p. 113[200] |

162 How utterly incomprehensible that if meteoric stones simply pitched from moon, that the metals should be those which have magnetic properties.

Study well products of Solfatarias. some general laws. association of lead & silver. Sulp. of Barytes: Fluoric. Barytes: — |

163e Humboldt. New Spain. Vol III. p. 130[201] Metals in Mexico rarely in secondary always in primitive & transition; the latter rarely appear in central Cordillera. particularly between 18° & 22° N. = formations of amph: porphyry. greenstone[,] amygdaloid. basalt & other trap cover it to great thickness. = Coast of Acapulco granitic rock. — in parts of table granits & gneiss with gold veins visible: — "Porphyries of Mexico may be considered for most parts as rock eminently rich in mines of gold & silver." [p. 131][202] |

164e The above porphyries characterized by no quartz & amphibole frequently only vitreous felspar: = gold veins in a phonolitic porphyry. = several parts of N. Spain great analogy to Hungary. = Veins of Zimapan offer zeolite. stilbite. grammalite. pyenite. native sulphur.. fluor spar. bayte. asbestos garnets. — carb & chrom. of lead. orpiment. chrysoprase. opal: —[203]

Veins in Limestone & Grauwacke: Silver appears far more abundant in the upper limestone, which H. calls by several secondary names[204] |

165e ⌈Study Hoffmans account of steam acting on trachytes. also Azores. We here have case of such vapours washing a rock[205]⌉ Veins concretionary; concretions determined by fissures as in septaria. (& Chiloe case, at least corelation) — Galapagos vein. vein of secretion. — metallic veins follow mountain chain. there after NW ⟨W⟩. — ⌈same chemical laws as in concretions perhaps makes intersections richest — Humboldt has urged phenomena in veins, chemical affinities like in composed rock.[206] granites syenite⌉ ⌈strangling &c of veins can only be accounted for by concretionary action, conjoined with other⌉ ⌈state simplest case. concretions of clay iron stone; iron pyrite in a fossil⌉ Insist strongly on the grand fact of Volcanic & non Volcanic. Then Solfataras. ⌈Mem: Micaceous iron ore.⌉

N.B. To show how metals may be transported by complicated chemical law & steam of salts, quite curious case of oxided Iron by Mitterschlich. Vol. II Journal of Nat. & Geograph Siciences? —[207] |

166e H says in Potosi the silver is contained in a primitive slate, covered by a clayey porphyry, containing grenats. In Peru. on other hand, mine of Gualgayoc or Chota & Pasco in "alpine limestone" = "The wealth of the veins in most part totally independent of the nature of the beds they intersect". = In the Guatemala part. (& Chiloe do) no veins discovered. Humboldt suggests covered up by volcanic rocks.[208] |

167e St Helena has been slightly broken up, & has there not been vein ⌈of iron⌉ discovered? —

Klaproth analysed silver ores from Peru consisted of native silver. & brown oxide of Iron in Mexico. sulphuretted silver, arsenical grey copper, and antimony, horn silver, black silver & red silver, do not name native silver because not very abundant. — muriated silver. which is so rare in Europe. common there accompanied by molybdated lead & ⌈argentiferous lead⌉; sulfated Barytes very ⌈un⌉common in Mexico. Fluor spar only in certain mines.[209] |

168e ⌈Vol. III⌉ "In general it is observed both in Mexico & Peru, that those oxidated masses of iron which contain silver are peculiar to that part of the veins, nearest to the surface of the earth." —

p. 156.[210] Mines of Batopilas in New Biscay, "Nature exhibits the same minerals there, that are found in the veins of Kongsberg in Norway. — namely dendritic silver intersecting carbonate of lime —[211] native silver in Mexico |

169e is always accompanied by Sulp. silver sometimes by selenite. —[212] in New Spain, contrary to Europe argentiferous lead not abundant. =[213] considerable quantity of silver procured from martial pyrites; great blocks of pure silver not common in ⟨S.⟩ America: In all climates distribution of silver [in veins] very unequal, sometimes disseminated sometimes concentrated: wonderful quantity of pure silver in S. America.[214] |

170e Geology of Guanuaxuato. — Clay slate. passing into talcose & chloritic slate. with beds of syenite & serpentine dipping to SW at 45° to 50° — covered by conformable greenstone porphyrys & phonolites do. amphibole quartz & mica very rare. —[215] ancient freestone & breccia is the same with that on surface of plains of Amazon, no relation — there is more modern breccia, chiefly owing to destruction of porphyries. whereas other to ancient rock. — this N° 2. superimposed on N° 1. even No. 2. might be mistaken for Porphyry |

171e above ancient freestone, limestone & ⟨many⟩ [other secondary] rocks.[216]

Vein traverses both Clay slate, Porphyry North 52 W, & is nearly the same with that of the veta grande of Zacatecas, & veins of Tasco & Moran — of Guanaxuato to SW. with respect to latter doubts whether bed or vein (very like that of Spital of Schemnitz in Hungary.) Humboldt says fragments from roof & penetrating overlying beds tells the secret. —[217] p. 189. "The small ravins into which the valley of Marfil is divided, appear to have a decided influence on the richness of the veta madre of [continued on page 175] |

170e [misnumbered page]

[172e]

Dr D. remarks. bad conductor of Heat do of Electricity[218]

Does not iron, combined with nickel & cobalt (meteoric) resist oxidation? — Mem Sir W. P. stone[219]

It is clear to me, there are laws of solution & deposition under great pressure. (? fact!) unknown to us.

M. Chladni. — on meteoric Mexican stone. Journal des Mines 1809. No. 151. p. 79.[220] |

[misnumbering, no page 173]

174e Under name of Sagitta Triptera D'Orbigny has figured animal with setæ like my undescribed[.] p. 140. Flèche of Quoy et Gaimard. — D'Orbigny has described it with care to 3 species. I think I have much additional information[221] |

175e [continued from page 171] Guanaxuato, which has yielded the most metal, where the direction of ravins, and the slope of the mountains (flaqueza del Cerro) have been parallel to the direction & inclination of the vein". —[222]

at Zacatecas the veta grande has same direction as Guanax. — the other E & W. — veins richest not in ravins or along gentle slopes. but on the most elevated summits, where mountains most torn. — (?anticlinal line?). —[223]

Mines of Catorce [(Principal veins)] 25° to 30° to NE. vein of Moran 84° NE. of Real del Monte 85° to S. // Tasco 40° to NW (afterwards said to be [all with some exception] directed NW & SE).[224] |

176e [Vol III] Mexican Cordillera "immense variety of Porphyries which are destitute of quartz, & wh abound both in hornblend & vitreous felspar". — p. 215[225]

Same metal in Tasco vein in Mica Slate & overlying Limestone[226]

Balls of Silver ore occur in do veins.[227] At Huantajaia. Humboldt says, mur of Silv.[,] Sulph. of do.[,]galena[,]quartz, Carb. of Lime. accompany. — Ulloa has said silver in the highest & gold in the lowest. Humboldt states that some of the richest gold mines on ridge of Cordillera near Pataz, also at Gualgayoc. where many petrified shells[228] |

177e Bougainville says P 291. —

The Fuegians treat the "chefs d'œuvre de l[']industrie humaine, comme ils traitent les loix de la nature & ses phenomenes." —[229]

Ulloa's Voyage, Shell fish purple die, marvellous statements on, Vol I, P. 168. on coast of Guayaquil, same as Galapagos.[230]

no Hydrophobia at Quito. P 281. do do[231]

Australia, C. of Good Hope. — Azores Is^ds^ [nor at St Helena. —][232]

Humboldt. New Spain Vol. IV. [p. 58.] At Acapulco earthquakes are recognized as coming from three directions. from W. NW & S. — last to Seaward[233] |

178 partaking of the character of a Araucarian tribe, with point affin of yew & intermediate[234]

Puncture one animal with recent dead body of other. & see if same effects, as with man

Does Indian rubber & black lead unite chemically like grease & mercury |

179 [blank]

180 N.B. P. 73. General reflections on the geology of the world

P. 14. } 91 } gradual shoaling of coasts

93 action of sea on coast.

27. Bahama Is^d^ |

181 De Lucs travels[235]

Beauforts Karamania[236]

Capt. Ross.[237] & Scoresby[238] deep soundings

Gilbert Farquhar Mathison travels Brazil. Peru. Sandwich [Isd][239]

Mawes travels down the Brazil. —[240]

Did Melaspena publish his travels?[241] |

Bellinghausen in 1819[242]

Kotzebue 1816[243] |

inside back cover

Constant log always additive to convert French Toise into English [ft.] 0.8058372

French metre into English ft. 0.5159929

	Toises	Pieds		
Myriametre =	5130.,	4.	5 inches	
Kilometre	513.,	0.	5	
Hectometre	51.	1.	10	
Metre		3.	0.	11 lines
Decimetre			3.8	
Centimetre				4.4

[C. Darwin] |

back
cover

R.N.

Range of Sharks
[Nothing For any Purpose]

The back cover of the Red Notebook, labelled 'R.N.', with the additional notations 'Range of Sharks' and 'Nothing For any Purpose'.

Editor's Notes

[1] Jean François d'Aubuisson de Voisins, *Traité de géognosie*. 2 vols. (Strasbourg, 1819).

[2] Juan Ignacio Molina, *Compendio de la historia geografica...del reyno de Chile* (Madrid, 1788), vol. 1.

[3] Sir Charles Lyell, F.R.S. (1797–1875), prominent British geologist, twice president of the Geological Society of London (1835–1837, 1849–1851), and author of the *Principles of Geology*. 3 vols. (London, 1830, 1832, 1833). This work exercised a formative influence on the development of geology as a science in the nineteenth century and on the career of Charles Darwin, F.R.S. (1809–1882). This entry in the notebook is in light brown ink.

[4] Lyell, *Principles of Geology*. This entry is in light brown ink, and written over the immediately preceding series of dates. The dates pertain to the departure of H.M.S. *Beagle* from England. The *Beagle* sailed from England Tuesday 27 December 1831. The ship encountered heavy seas, caused by gales elsewhere, on Thursday 29 December 1831. For Darwin's description of the *Beagle*'s departure see his letter to his father of 8 February–1 March 1832 in Nora Barlow, ed., *Charles Darwin and the Voyage of the Beagle* (London, 1945), p. 52. Also see N. Barlow, ed., *Charles Darwin's Diary of the Voyage of H.M.S. Beagle* (Cambridge, 1933), pp. 18–19. Darwin could have recorded the date of the *Beagle*'s departure in this notebook at any time during the voyage.

[5] The probable stimulus for this passage was Christian Gottfried Ehrenberg, 'On the Origin of Organic Matter from simple Perceptible Matter, and on Organic Molecules and Atoms; together with some Remarks on the Power of Vision of the Human Eye' in Richard Taylor, ed., *Scientific Memoirs* (London, 1837), vol. 1, pp. 555–576. This entry is in light brown ink, indicating a later dating than the original entries on this page.

[6] Jacques Julien Houton de Labillardière, *Relation du voyage à la recherche de La Pérouse...1791*–[1794] (Paris, 1800), vol. 1, p. 287: "Je revis le fucus que j'avois auparavant rencontré tout près de la Nouvelle-Guinée; il ressemble à de l'étoupe très-fine coupée par petis morceaux longs d'environ trois centimètres: ce sont des filamens aussi fins que des cheveux. On les voyoit souvent réunis en faisceaux, et si nombreux qu'ils ternissoient l'eau de la rade."

[7] John Stevens Henslow, 'Geological Description of Anglesea', *Transactions of the Cambridge Philosophical Society*, vol. 1 (1821–1822), p. 379: "The major axis of

some of the larger nodules is two feet and a half, and the minor one foot and a half; and the conical structure extends to the depth of three or four inches. The direction of the longer axis is placed parallel to the schistose laminae, which pass round the nodules."

[8] William Fitton, 'Geology' in Phillip P. King, *Narrative of a Survey of the Intertropical and Western Coasts of Australia Performed between the Years 1818 and 1822* (London, 1827), vol. 2, p. 585: "The Epidote of Port Warrender and Careening Bay, affords an additional proof of the general distribution of that mineral; which though perhaps it may not constitute large masses, seems to be of more frequent occurrence as a component of rocks than has hitherto been supposed."

[9] Henslow, 'Geological Description of Anglesea', p. 403: "Carbonate of lime is very generally disseminated through every part [of the Plas-Newydd dike]."

[10] Henslow, 'Geological Description of Anglesea', p. 417: "The most interesting phenomena exhibited by this dyke, are the various changes which it assumes in its mineral character."

[11] Henslow, 'Geological Description of Anglesea', p. 434: "Through this dyke there run several veins of quartz, which also abound in the surrounding rock, a fact which I do not recollect witnessing in any other dyke in Anglesea." Also p. 419: "At its [the dyke's] Northern termination the trap has been removed by the continued action of the sea, and its original walls, composed of quartz rock, form a small bay about eighty feet wide."

[12] Henslow, 'Geological Description of Anglesea', p. 375: "As the limestone passes into the schist [at Gwalchmai], it assumes a fissile character, and scales of chlorite are dispersed over the natural fractures."

[13] Henslow, 'Geological Description of Anglesea', p. 432: "The whole [mass of trap] assumes a greenish tinge, but the colouring substance does not appear to be of a very crystalline nature, and is probably chlorite."

[14] See William Dampier, *A New Voyage round the World* (4th ed.; London, 1698–1703), vol. 2 [1699], part 3 subtitled: *A Discourse of Trade-Winds, Breezes, Storms, Seasons of the Year, Tides and Currents of the Torrid Zone throughout the World: With an Account of Natal in Africk, its Product, Negro's, &c.*

[15] Constantin François Volney, *Voyage en Syrie et en Égypte...1783–1785* (2nd ed. rev.; Paris, 1787), vol. 1, chapter 20 the section entitled 'Des vents', and chapter 21 entitled 'Considérations sur les phénomènes des vents, des nuages, des pluies, des brouillards et du tonnerre'.

[16] Dampier, *A New Voyage round the World*, vol. 3 [1703], p. 125: "Of the Sharks we caught a great many, which our Men eat very favourily. Among them we

caught one which was 11 Foot long." I have not found an edition of this work which fits Darwin's page citation exactly. (Ed.)

[17] Dampier, *A New Voyage round the World*, vol. 3 [1703], pp. 125–126: "Its Maw was like a Leather Sack, very thick, and so tough that a sharp Knife could scarce cut it: In which we found the Head and Boans of a *Hippopotomus*; the hairy Lips of which were still sound and not putrified, and the Jaw was also firm, out of which we pluckt a great many Teeth, 2 of them 8 Inches long, and as big as a Mans Thumb, small at one end, and a little crooked; the rest not above half so long. The Maw was full of Jelly which stank extreamly:...'Twas the 7th of *August* when we came into *Shark*'s Bay;..."

[18] Dampier, *A New Voyage round the World*, vol. 3 [1703], p. 114: "At about 30 Leagues distance [from the Abrolhos shoals] we began to see some Scutle-bones floating on the Water; and drawing still nigher the Land we saw greater quantities of them." Also p. 115: "The 30th of *July*, being still nearer the Land, we saw abundance of Scutle-bones and Sea-weed, more Tokens that we were not far from it;..."

[19] Capt. Samuel P. Henry (1800–1852), author of *Sailing Directions for Entering the Ports of Tahiti and Moorea* (London, 1852); personal communication. Darwin met Capt. Henry and his father, a missionary, at Tahiti. See Robert Fitzroy, ed. *Narrative of the Surveying Voyages of His Majesty's Ships Adventure and Beagle...1826–1836* (London, 1839), vol. 2, pp. 524, 546, 615; and John Williams, *A Narrative of Missionary Enterprises in the South Sea Islands* (London, 1837), p. 471.

[20] Labillardière, *Relation du voyage à la recherche de La Pérouse*, vol. 1, p. 394: "L'îlot sur lequel nous étions est composé d'un beau granit, où le quartz, le feld-spath et le mica dominent;..." and, p. 395, "La partie occidentale de cet îlot offre, dans un des points les plus élevés un plateau de pierre calcaire...."

[21] Dampier, *A New Voyage round the World*, vol. 3 [1703], p. 151: "The Land hereabouts was much like that part of *New Holland* that I formerly described....'tis low, but seemingly barricado'd with a long Chain of Sand-hills to the Sea, that let's nothing be seen of what is farther within Land."

[22] Jean François Galaup de La Pérouse, *A Voyage round the World Performed in the Years 1785, 1786, 1787, and 1788* (London, 1799), vol. 2, p. 179: "From Norfolk Island, till we got sight of Botany Bay, we sounded every evening with a line of two hundred fathoms, but we found no bottom till we were within eight leagues of the coast, when we had ninety fathoms of water."

[23] Frederick William Beechey, *Narrative of a Voyage to the Pacific and Beering's Strait...1825, 26, 27, 28* (Philadelphia, 1832). See note 59.

[24] This entry is in light brown ink.

[25] The sequence of points on this list runs from south to north along the Brazilian coastline. Place-names and latitudes were checked against British Admiralty charts of the period. Useful in this regard was the *Index to Admiralty Published Charts*, (London, 1874) published by the Hydrographic Office. A bar with a dot over a number indicates that no bottom was found at that depth. Undoubtedly Darwin compiled this list from information available to him aboard ship.

[26] This entry is in light brown ink. As a later addition it would appear to be a correction to the two figures immediately following, although only the '60' is actually cancelled. The sense of the passage would be that at 18–20 leagues from shore no bottom was found at 120 fathoms.

[27] Probably Robert Fitzroy, F.R.S. (1805–1865), Captain of H.M.S. *Beagle* during its surveying voyage of 1831–1836, later vice-admiral in the navy and a meteorologist of considerable repute. It was with Fitzroy's assent that Charles Darwin became the *Beagle*'s naturalist. For reference to Fitzroy's account of the *Beagle*'s voyage see note 19.

[28] Mrs Power, presumably a resident of Port Louis, Mauritius; personal communication. Mrs Power is not mentioned otherwise in Darwin's notes.

[29] The shipwrecked crew of the H.M.S. *Wager* identified their position as 47° 00′ S., 81° 40′ W. Capt. Fitzroy recalculated the probable position of the ship as 47° 39′ 30″ S., 75° 06′ 30″ W. See John Bulkeley and John Cummins, *A Voyage to the South-Seas...1740–1* (London, 1743), p. 48; and Fitzroy, ed., *Narrative of the Surveying Voyages of His Majesty's Ships Adventure and Beagle...1826–1836*, appendix to vol. 2, p. 78. The earthquakes of August 25, 1741 experienced by the shipwrecked crew of the *Wager* were described as "four great Earthquakes, three of which were very terrible; notwithstanding the violent Shocks and Tremblings of the Earth, we find no Ground shifted. Hard Gales of Wind at North, with heavy Showers of Rain." (Bulkeley and Cummins, p. 70.) Also see *JR*, p. 287.

[30] Henry Thomas De La Beche, *A Geological Manual* (London, 1831), sections 5–10.

[31] De La Beche, *A Geological Manual*, section 3, 'Erratic Blocks and Gravel'. In his treatment of the subject De La Beche did not discuss the shapes of individual pieces of gravel.

[32] Robert Were Fox, 'On the electro-magnetic properties of metalliferous veins in the mines of Cornwall', *Philosophical Transactions of the Royal Society of London*, vol. 120 (1830), pp. 399–414.

[33] Joseph Carne, 'On the relative age of the Veins of Cornwall', *Transactions of the Royal Geological Society of Cornwall*, vol. 2 (1822), pp. 49–128.

[34] Baron Albin-Reine Roussin (1781–1854), French naval commander and later admiral, member of the Académie des Sciences, did not write a general account of the hydrographical expedition he led in 1819–1820 to South America. Darwin was already familiar with the technical publication stemming from the voyage, Roussin's *Le Pilote du Brésil* (Paris, 1826).

[35] William D. Conybeare and William Phillips, *Outlines of the Geology of England and Wales* (London, 1822), p. xii: "...one instance of a bone penetrated by silex has occurred to the author, on the beach at Reculver. The calcareous substance of shells, echinites, encrinites, corals, &c. in its slightest change seems only to have lost its colouring matter and gelatine; next they become impregnated with the mineral matrix in which they lie, especially if that matrix be calcareous; hence they become much more compact; often at the same time their original calcareous matter undergoes a change of internal structure, assuming a crystalline form, and in some cases, viz. asteriæ, encrinites, and echinites, a calcareous spar of very peculiar character results, of an opaque cream colour:..."

[36] Conybeare and Phillips, *Outlines of the Geology of England and Wales*, p. xv: "These consolidated gravel beds are called conglomerates, breccias, or pudding-stones; we find them among the transition rocks, in the old red sandstone, in the millstone-grit and coal-grits, in the lower members of the new red sandstone, in the sand strata beneath the chalk, and in the gravel beds associated with the plastic clay, and interposed between the chalk and great London clay."

[37] Despite the faulty citation the reference is certainly to Joseph Jackson Lister, 'Some Observations on the Structure and Functions of tubular and cellular Polypi, and of Ascidiæ', *Philosophical Transactions of the Royal Society of London*, vol. 126 (1834), pp. 365–388.

[38] Alexander von Humboldt, *Personal Narrative of Travels to the Equinoctial Regions of the New Continent...1799–1804* (London, 1829), vol. 7, p. 56: "Farther south, towards Regla and Guanabacoa [to the east of Havana], the syenite disappears, and the whole soil is covered with serpentine, rising in hills from 30 to 40 toises high, and running from east to west." Darwin's copy of Humboldt's *Personal Narrative* is inscribed, "J. S. Henslow to his friend C. Darwin on his departure from England upon a voyage round the World. 21 Sept 1831." It consists of vols. 1–2, 3rd ed.; vol. 3, 2nd ed.; vols. 4, 5, 6, 7, 1st ed. (London, 1819–1829). Alexander von Humboldt (1769–1859), a member of all major scientific academies, was the foremost scientific traveller of his day and a principal contributor to the science of geography.

[39] ['Proteus'], 'The Bahama Islands', *United Service Journal and Naval and Military Magazine*, vol. 3 (1834), p. 215: "[New Providence] is more hilly than most of the islands, the surface being composed of rock and sand intermixed with sea shells." Also see pp. 216 and 226 for mention of the banks.

[40] Sir John F. W. Herschel, F.R.S. (1792–1871), distinguished English astronomer and man of science; presumably personal communication. Darwin met Herschel—"the most memorable event which, for a long period, I have had the good fortune to enjoy"—sometime between 8–15 June 1836 during the *Beagle*'s call at the Cape of Good Hope where Herschel was living, being then engaged in his four-year study of stars visible in the southern hemisphere. See *Diary*, p. 409. Months before, Herschel had described his new notion of the cause of volcanic action in a letter to Charles Lyell dated 20 February 1836. Probably he repeated the same explanation to Darwin in June. Herschel's letter to Lyell has been published by Walter F. Cannon in 'The Impact of Uniformitarianism', *Proceedings of the American Philosophical Society*, vol. 105 (1961), pp. 301–314. See, for example, Herschel's summary comment to Lyell on p. 310: "I don't know whether I have made clear to you my notions about the effects of the removal of matter from...above to below the sea.—1st it produces mechanical subversion of the *equilibrium of pressure*.—2dly it also, & by a different process (as above explained at large) produces a subversion of the equilibrium of temperature. The last is the most important. It *must be an excessively slow process*. & it will depend 1st on the depth of matter deposited.—2^{d} on the quantity of water retained by it under the great squeeze it has got—3dly on the tenacity of the incumbent mass—whether the influx of caloric from below— which MUST TAKE PLACE acting on that water, shall either heave up the whole mass, as *a new continent*—or shall crack *it* & escape as a submarine volcano—or shall be suppressed until the mere weight of the continually accumulating mass breaks its lateral supports at or near the coast lines & opens there a chain of volcanoes."

[41] Sir Andrew Smith, F.R.S. (1797–1872), English army medical doctor and zoologist, later director-general army medical department; personal communication. Darwin's *Diary*, p. 409, records for 8–15 June 1836: "During these days I became acquainted with several very pleasant people. With Dr A. Smith who has lately returned from his most interesting expedition to beyond the Tropic, I took some long geological rambles." On his return to England in 1837 Smith began work on his *Illustrations of the Zoology of South Africa*. 5 parts. (London, 1838–1849).

[42] The meaning of this entry is obscure. The H.M.S. *Chanticleer* did not stop at Pernambuco [Recife] during its 1828–1831 voyage, nor was Pernambuco on the *Beagle*'s itinerary in June of 1836, when this entry was presumably made. In the narrative from the *Chanticleer*'s voyage, however, there are passages which describe decomposing granitic rock at Rio de Janeiro, and refer to what seem to be related formations at Para [Belém] and Maranham [São Luís]. Given Darwin's apparent uncertainty in this entry about location, as indicated by his two cancellations, it may have been these passages which he had in mind. See William H. B. Webster, *Narrative of a Voyage to the Southern Atlantic Ocean...1828–1830* [sic] (London, 1834), vol. 1, pp. 52–53: "The country about Rio in a geological point of view has large claims to attention. Granite and gneiss are the prevailing formation....The rocks in some

parts are decomposed into sand and petunse; the sand having been carried down into the plains, while the petunse remains, and forms extensive beds of porcelain clay admirably adapted for the use of the potter. The lower parts of the granite hills were found chiefly in this condition; the granite having crumbled into micaceous sand and greasy unctuous clay." Also vol. 2, pp. 367: "The geology of Para will detain us a very little while; as there is very little variety or novelty. Precisely the same materials are found here as at Maranham, so that it would be impossible to distinguish them. It is a rare and unusual circumstance to find such a striking coincidence, in two different places. The soil upon which the city stands is of clay and sand. The beds of clay are very extensive, and frequently thirty or forty feet deep. There is scarcely any rock, and that only in particular and isolated masses; it is a coarse dark iron sand-stone, with numerous particles of quartz in it.... This dark iron sand-stone, with fragments of white quartz, is observable at Maranham, and is the predominant formation at St. Paul's, a little to the southward of Rio."

[43] The clipping, entitled 'Earthquake at Sea' is from the *Carmarthen Journal*, 3 April 1835. The story was reprinted verbatim from *The Times* (London), 28 March 1835, p. 5, with the unfortunate error of a lost digit in the quotation of the ship's latitude. The ship's coordinates as given in *The Times* were 18° 47′ N., 61° 22′ W., which would place the ship in the Atlantic Ocean to the northeast of the Leeward Islands, rather than, as in the incorrectly printed version, in Venezuela.

[44] The H.M.S. *Challenger* ran aground on the Chilean shore at Punta Morguilla [Point Molguilla] (37° 46′ S., 73° 40′ W.) on 19 May 1835. See Fitzroy, ed., *Narrative of the Surveying Voyages of His Majesty's Ships Adventure and Beagle*, vol. 2, pp. 451–456. Capt. Fitzroy led the party which rescued the *Challenger*'s crew.

[45] This paragraph is double scored in the left margin with brown ink.

[46] In this series of place names the locations of Guacho and Washington are uncertain. There is presently a Quebrado del Guacho, a small stream, at 33° 58′ S., 71° 09′ W. in Chile, and a Cerro Guacho, a mountain, nearby. 'Washington' may refer to the Canal Washington at 55° 40′ S., 67° 33′ W. in Tierra del Fuego.

[47] Fitton, 'Geology', in King, *Narrative of A Survey of the Intertropical and Western Coasts of Australia*, vol. 2, p. 604: "The tendency of all this evidence is somewhat in favour of a general parallelism in the range of the strata,—and perhaps of the existence of primary ranges of mountains on the east of Australia in general, from the coast about Cape Weymouth to the shore between Spencer's Gulf and Cape Howe." And on p. 605: "If...future researches should confirm the indications above mentioned, a new case will be supplied in support of the principle long since advanced by Mr. Michell which appears (whatever theory be formed to explain it,) to be established by geological observation in so many other parts of the world,—that the outcrop of the inclined beds, throughout the stratified portion of the globe, is every

where parallel to the longer ridges of mountains,—towards which, also, the elevation of the strata is directed."

[48] Charles Daubeny, *A Description of Active and Extinct Volcanos* (London, 1826), p. 24: "It [a formation at the hill of Mouton] should be noticed, as one of the few localities in Auvergne where pumice is to be found, which seems the more remarkable, as this substance is a common product of that class of volcanos, which consists of trachyte."

[49] This entry is in light brown ink. The back of page 1, of Darwin's geological notes on New Zealand is fol. 802 verso in the Darwin MSS, Cambridge University Library, vol. 37 (ii). The page contains a sketch of the silhouette of an island in the Bay of Islands, New Zealand. Darwin noted that at high water the island had the figure of a hill and at low water the figure of a hill surrounded by a level ledge of naked rock. He associated the formation of the ledge with the action of the tides. This page in Darwin's geological notes also contains a cross-reference to 'R.N.' page 38.

[50] See *GSA*, pp. 25–26, for the published version of this description of the origin of the cliffs at St Helena.

[51] Daubeny, *Volcanos*, reference uncertain, possibly to the author's representation of Humboldt's 'unpublished' views on pp. 345–351. 'Daubeny' is written in light brown ink.

[52] Humboldt, *Personal Narrative*, vol. 1, p. 171: "The Peak of Teneriffe, and Cotopaxi, on the contrary, are of very different construction. At their summit a circular wall surrounds the crater; which wall, at a distance, has the appearance of a small cylinder placed on a truncated cone." Also, with respect to the peak of Teneriffe, on p. 176: "The wall of compact lava which forms the enclosure of the Caldera, is snow white at it's surface....When we break these lavas, which might be taken at some distance for calcareous stone, we find in them a blackish brown nucleus. Porphyry with basis of pitch stone is whitened externally by the slow action of the vapors of sulphurous acid gas."

[53] Humboldt, *Personal Narrative*, vol. 1, pp. 219–232.

[54] Daubeny, *Volcanos*, p. 349. Not easily summarized, see note 51.

[55] Daubeny, *Volcanos*, p. 361: "Humboldt gives us the following series of phænomena, which presented themselves on the American Hemisphere between the years 1796 and 97, as well as between 1811 and 1812.

1796.—September 27.	Eruption in the West India Islands; volcano of Guadaloupe in activity.
.......November ...	The volcano of Pasto begins to emit smoke.
.......December 14.	Destruction of Cumana by earthquake.

1797.—February 4...	Destruction of Riobamba by earthquake.
1811.—January 30...	Appearance of Sabrina Island in the Azores. It increases particularly on the 15th of June.
.......May	Beginning of the earthquakes in the Island of St. Vincent, which lasted till May, 1812.
.......December 16.	Beginning of the commotions in the valley of the Mississippi and Ohio, which lasted till 1813.
.......December ...	Earthquake at Carracas.
1812.—March 26....	Destruction of Caraccas; earthquakes which continued till 1813.
.......April 30.....	Eruption of the volcano in St. Vincents'; and the same day subterranean noises at Caraccas, and on the banks of the Apure."

[56] Daubeny, *Volcanos*, pp. 382–383: "With regard to the mineralogical characters of lava, I shall appeal to the authority of [Leopold] Von Buch....Almost all lavas he conceives to be a modification of trachyte, consisting essentially of felspar united with titaniferous iron, to which they owe their colour and their power of attracting iron.... This felspar is derived immediately from trachyte, that being the rock which directly surrounds the focus of the volcanic action; for if we examine the strata that successively present themselves on the sides of a crater, we are sure to find that the lowest in the series is trachyte, from which is derived by fusion the obsidian, as is the case at Teneriffe." Leopold von Buch (1774–1853), German geologist and mineralogist, a member of the Royal Academy of Berlin, was distinguished for the versatility of his interests in geology and for the high quality of his extensive field work.

[57] Daubeny, *Volcanos*, p. 386: "...in the collection of Dr. Thomson, now in the Museum of Edinburgh, there is said to be a fragment of lava enclosing a real granite, which is composed of reddish felspar with a pearly lustre like adularia, of quartz, mica, hornblende, and lazulite. I have likewise seen among the specimens from the Ponza Islands,...a piece of granite, or perhaps rather of a syenitic rock,...found in the midst of the trachyte from this locality. But the most interesting fact perhaps of this description, is...the presence of a mass of granite containing tin-stone, enveloped in the midst of a stream of lava from Mount Ætna....It may be remarked, that these specimens of granitic rocks have, in general, a degree of brittleness, which accords very well with the notion of their exposure to fire."

[58] Lyell, *Principles of Geology*, vol. 1, p. 318 refers to Java "where there are thirty-eight large volcanic mountains, many of which continually discharge smoke and sulphureous vapours." This entry is written in light brown ink.

[59] Beechey, *Narrative of a Voyage to the Pacific and Beering*'s *Strait*, p. 209: "In latitude 60° 47′ N. we noticed a change in the colour of the water, and on sounding found fifty-four fathoms, soft blue clay. From that time until we took our final departure from this sea the bottom was always within reach of our common lines. The water shoaled so gradually that at midnight on the 16th, after having run a hundred and fifty miles, we had thirty-one fathoms." P. 211: "We soon lost sight of every distant object, and directed our course along the land [St Lawrence Island], trying the depth of water occasionally. The bottom was tolerably even; but we decreased the soundings to nine fathoms, about four miles off the western point, and changed the ground from fine sand, to stones and shingle. When we had passed the wedged-shaped cliff at the north-western point of the island, the soundings again deepened, and changed to sand, as at first.... [Zoological specimens were procured] in seventeen fathoms over a muddy bottom, several leagues from the island." P. 212–213: "In our passage from the St. Lawrence Island to this situation, the depth of the sea increased a little, until to the northward of King's Island, after which it began to decrease; but in the vicinity of the Diomede Islands, where the strait became narrowed, it again deepened, and continued between twenty-five and twenty-seven fathoms. The bottom, until close to the Diomedes, was composed of fine sand, but near them it changed to course stones and gravel, as at St. Lawrence Island...." P. 213: "Near the Asiatic coast we had a sandy bottom, but, in crossing over the [Beering's] strait, it changed to mud, until well over on the American side, where we passed a tongue of sand and stones in twelve fathoms which, in all probability, was the extremity of a shoal, on which the ship was nearly lost the succeeding year. After crossing it, the water deepened, and the bottom again changed to mud, and we had ten and a half fathoms within two and a half miles of the coast." P. 444: "In this parallel [61° 58′ N] the nearest point of land bearing N. 74° W. true, thirteen miles, the depth of water was 26 fathoms; and it increased gradually as we receded from the coast.... We made the land [St Lawrence Island] about the same place we had done the preceding year, stood along it to the northward, and passed its N.W. extreme, at two miles and a half distance, in 15 fathoms water, over a bottom of stones and shells, which soon changed again to sand and mud.... On the after-noon of the 2d we... anchored off Point Rodney... in seven fathoms, three miles from the land...."

[60] In 'at' an upper case 'a' has been superimposed on a lower case 'a'.

[61] The 'turn over' indicates that the entry continues on the next page. The entire paragraph at the bottom of page 45e is scored for emphasis in light brown ink.

[62] The quotation is from Daubeny, *Volcanos*, p. 402 which summarizes the argument presented in Conybeare and Phillips, *Outlines of the Geology of England and Wales*, p. xx.

[63] The question mark is written in light brown ink.

[64] See note 62.

[65] See Daubeny, *Volcanos*, p. 438 for the following note: "Cet endroit [near the Red Sea] recouvert de sable, environné de rochers bas en forme d'amphitheatre, offre une pente rapide vers la mer dont il est eloignè d'un demi mille, et peur avoir trois cent pieds de hauteur sur quatre-vingts de largeur. On lui a donné la nom de Cloche, parcequ'il rend des sons, non comme faisait autrefois la statue de Memnon, au lever du soleil, mais à toute heure du jour et de la nuit et dans toutes les saisons. La premiere fois qu'y alla M. Gray, il entendit au bout d'un quart d'heure un son doux et continu sous ses pieds, son, qui en augmentant ressembla à celui d'une clocha qu'on frappe, el qui devient si fort en cinq minutes, qu'il fit detacher du sable, et effraya les chamaux jusqu'â les mettre en fureur." Also see *JR*, p. 441.

[66] Volney, *Voyage en Syrie et en Égypte*, vol. 1, p. 351 with reference to the deserts of Syria: "Presque toujours également nue, la terre n'offre que des plantes ligneuses clair-semées, et des buissons épars, dont la solitude n'est que rarement troublée par des gazelles, des lièvres, des sauterelles et des rats."

[67] Lyell, *Principles of Geology*, vol. 2, chap. 11 bears the following summary heading: "Theory of the successive extinction of species consistent with their limited geographical distribution—The discordance in the opinions of botanists respecting the centres from which plants have been diffused may arise from changes in physical geography subsequent to the origin of living species—Whether there are grounds for inferring that the loss from time to time of certain animals and plants is compensated by the introduction of new species?—Whether any evidence of such new creations could be expected within the historical era, even if they had been as frequent as cases of extinction?—The question whether the existing species have been created in succession can only be decided by reference to geological monuments."

[68] In this case 'from' is written over 'the'.

[69] John Miers, *Travels in Chile and La Plata* (London, 1826), vol. 1, p. 77: "About two miles to the eastward of Barranquitos [32° 35′ S., 64° 20′ W.] I picked out of the sand a small fragment of quartz, about half the size of a hazel nut. This was the first pebble or stone of any sort I had seen since I left Buenos Ayres."

[70] Charles Marie de La Condamine, *A Succinct Abridgment of a Voyage Made within the Inland Parts of South-America* (London, 1747), p. 24: "Below *Borja*, even for four or five hundred leagues, a stone, even a single flint, is as great a rarity as a diamond would be. The savages of those countries don't know what a stone is, and have not even any notion of it. It is diversion enough to see some of them, when they come to *Borja*, and first meet with stones, express their admiration of them by signs, and be eager to pick them up; loading themselves therewith, as with a valuable merchandize; and soon after despise and throw them away, when they perceive them to be so common." See *JR*, p. 289.

[71] The phrase 'Carnatic | It has been common practice of geologists' appears in very small handwriting in light brown ink, which indicates that it was written some time after the other entries on pages 56e–57. Fortunately, however, despite the fragmentary nature of the entry, there exists a reference in Darwin's notes from the voyage, again by way of addition made in light brown ink, which identifies the use of 'Carnatic' in this context. See Darwin MSS, Cambridge University Library, vol. 33, fol. 115 verso, for citation of the following reference. James Allardyce, 'On the Granitic Formation, and direction of the Primary Mountain Chains, of Southern India', *Madras Journal of Literature and Science*, vol. 4 (1836), pp. 332–333: "It has been remarked that granite in America is found at a much lower level than in Europe: this is also the case throughout the south of India, by granite—meaning always granitic rocks; for a regularly crystallized compound of quartz, felspar and mica, is not to be expected. The Carnatic, and several other similar tracts, occurring along both coasts, are, as granitic plains, surprisingly level: the slight tertiary diluvium with which they are covered, cannot be considered as a principal cause of this uniformity, for the rock itself is everywhere found near the surface: every appearance here indicates the granitic formation has at one time been a great deal more flat than it is generally understood to have been."

[72] Lyell, *Principles of Geology*, vol. 3, p. 84: "It is clear, from what we before said of the gradual manner in which the principal cone [of Etna] increases, partly by streams of lava and showers of volcanic ashes ejected from the summit, partly by the throwing up of minor hills and the issuing of lava-currents on the flanks of the mountain, that the whole cone must consist of a series of cones enveloping others, the regularity of each being only interrupted by the interference of the lateral volcanos."

[73] This question mark and a line of scoring alongside the preceding sentence are in light brown ink.

[74] 'Rapilli' was equivalent in meaning to 'lapilli'. See, for example, the use of 'rapilli' by Daubeny (*Volcanos*, p. 251) and Humboldt (*Personal Narrative*, vol. 1, p. 232).

[75] This entry is in light brown ink.

[76] An oval depression towards the eastern end of Ascension Island was described by the resident English marines as the cricket ground because "the bottom is smooth and perfectly horizontal." See Darwin MSS, Cambridge University Library, vol. 38(ii), fol. 941 verso.

[77] Lyell, *Principles of Geology*, vol. 3, p. 111 begins the section entitled "Sea-cliffs—proofs of successive elevation." Lyell's point is stated most succinctly on page 113 where he cites the testimony of another author writing on the alterations produced by the sea on calcareous rocks on the shores of Greece "that there are four or five

distinct ranges of ancient sea cliffs, one above the other, at various elevations in the Morea, which attest as many *successive* elevations of the country."

[78] In this passage Darwin would seem to be addressing Lyell's argument (*Principles of Geology*, vol. 3, p. 114) that "...a country that has been raised at a very remote period to a considerable height above the level of the sea, may present nearly the same external configuration as one that has been more recently uplifted to the same height."

[79] Lyell, *Principles of Geology*, vol. 3, p. 116: "...we have seen [for the newer Pliocene] that a stratified mass of solid limestone, attaining sometimes a thickness of eight hundred feet and upwards, has been gradually deposited at the bottom of the sea, the imbedded fossil shells and corallines being almost all of recent species. Yet these fossils are frequently in the state of mere casts, so that in appearance they correspond very closely to organic remains found in limestones of very ancient date."

[80] René Primevère Lesson and Prosper Garnot, *Voyage autour du monde...1822–1825. Zoologie* (Paris, 1826), vol. 1, part 1, p. 14: "...mais il est à remarquer que cette île vaste et composée de deux terres séparées par un détroit, quoique rapprochée de la Nouvelle-Hollande et par la même latitude, en diffère si complétement, qu'elles ne se ressemblent nullement dans leurs productions végétales. Toutefois la Nouvelle-Zélande, si riche en genres particuliers à son sol et peu connus, en a cependant d'indiens, tels que des piper, des olea, et une fougère réniforme qui existe, à ce qu'on assure, à l'île Maurice." Also p. 22: "Il est à remarquer qu'on ne connaît aucun quadrupède comme véritablement indigène de la Nouvelle-Zélande, excepté le rat, si abondamment répandu sur les îles de l'Océanie, comme sur presque l'univers entier."

[81] Molina, *Compendio de la historia geografica...del reyno de Chile*, vol. 1, p. 30: "La erupcion mas famosa de que tenemos noticia, fue la del volcan del monte de *Peteroa*, que el dia tres de Diciembre del ãno 1762 se abrió una nueva boca ó *cratéra*, hendiendo en dos partes un monte contiguo por espacio de muchas millas. El estrepito fue tan horrible, que se sintió en una gran parte del Reyno, pero no causó vibracion alguna sensible. Las cenizas y las lavas rellenaron todos los valles inmediatos, y aumentaron por dos dias las aguas del rio *Tingiririca*; y precipitandose un pedazo de monte sobre el gran rio *Lontué*, suspendió su corriente por espacio de diez dias, y estancadas las aguas, despues de haber formado una dilatada laguna que exîste en el dia, se abrieron por ultimo con violencia un nuevo camino, é inundaron todos aquellos campos." Darwin noted this passage in his own copy of the work with the remark, "P 30 — Piteron Earthquake caused lake & deluge — state of valleys." This entry is in light brown ink.

[82] Lyell, *Principles of Geology*, vol. 3, p. 124: "Towards the centre [of the dikes at Somma, the ancient cone of Vesuvius]...the rock is coarser grained, the component elements being in a far more crystalline state, while at the edge the lava is sometimes

vitreous and always finer grained. A thin parting band, approaching in its character to pitchstone, occasionally intervenes on the contact of the vertical dike and intersected beds. M. Necker mentions one of these at the place called Primo Monte, in the Atrio del Cavallo; I saw three or four others in different parts of the great escarpment."

[83] William F. W. Owen, *Narrative of Voyages to...Africa, Arabia, and Madagascar* (London, 1833), vol. 2, pp. 274–275: "[at Benguela]...the elephants were likewise common, but at present are scarce. A number of these animals had some time since entered the town in a body, to possess themselves of the wells, not being able to procure any water in the country. The inhabitants mustered, when a desperate conflict ensued, which terminated in the ultimate discomfiture of the invaders, but not until they had killed one man and wounded several others."

[84] Lyell, *Principles of Geology*, vol. 2, p. 189: "Thousands of carcasses of terrestrial animals are floated down every century into the sea, and, together with forests of drift-timber, are imbedded in subaqueous deposits, where their elements are imprisoned in solid strata...." Also p. 247: "...we see the putrid carcasses of dogs and cats, even in rivers, floating with considerable weights attached to them...."

[85] Claude Gay, 'Aperçu sur les recherches d'histoire naturelle faites dans l'Amérique du sud, et principalement dans le Chili, pendent les années 1830 et 1831', *Annales des sciences naturelles*, vol. 28 (1833), p. 371: "Ces contrées [Rio de Janeiro, Monte Video, Buenos Aires] m'offrirent aussi une assez belle collection d'insectes et plusieurs coquilles fluviatiles et marines, telles que des Mytilus, des Solens, des Ampullaires, etc., qui offraient ce phénomène digne de remarque, de vivre pêle-mêle dans les eaux simplement saumâtres." See *JR*, p. 24.

[86] De La Beche, *Geological Manual*, p. 73: "The Chesil Bank, connecting the Isle of Portland with the main land, is about sixteen miles long, and...the pebbles increase in size from west to east...The sea separates the Chesil Bank from the land for about half its length, so that, for about eight miles, it forms a shingle ridge in the sea. The effects of the waves, however, on either side are very unequal; on the western side the propelling and piling influence is considerable, while on the eastern, or that part between the bank and the main land, it is of trifling importance."

[87] Capt. Robert Fitzroy (note 27).

[88] Lyell, *Principles of Geology*, vol. 3, pp. 210–211: "The situation of this cliff [at Dax, France], is interesting, as marking one of the pauses which intervened between the successive movements of elevation whereby the marine tertiary strata of this country were upheaved to their present height, a pause which allowed time for the sea to advance and strip off the upper beds a,b, from the denuded clay c."

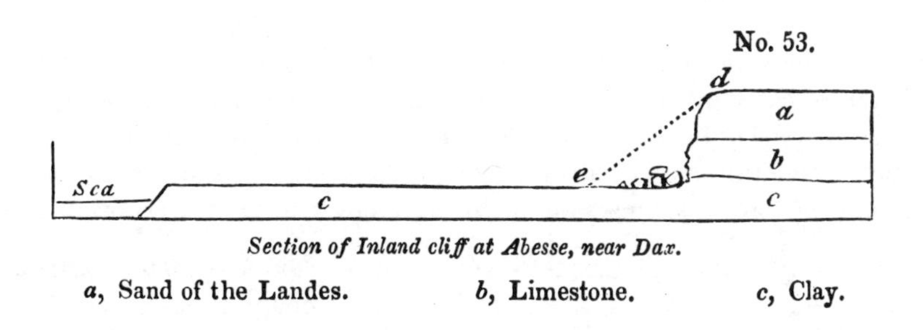

Section of Inland cliff at Abesse, near Dax.

a, Sand of the Landes. *b*, Limestone. *c*, Clay.

[89] Thomas Falkner, *A Description of Patagonia* (London, 1774), p. 51: "Being in the Vuulcan, below Cape St. Anthony, I was witness to a vast cloud of ashes being carried by the winds, and darkening the whole sky. It spread over great part of the jurisdiction of Buenos-Ayres, passed the River of Plata, and scattered it's contents on both sides of the river, in so much that the grass was covered with ashes. This was caused by the eruption of a volcano near Mendoza; the winds carrying the light ashes to the incredible distance of three hundred leagues or more."

[90] Lyell, *Principles of Geology*, vol. 3, p. 204: "Some of these bones [in certain strata in the basin of the Loire] have precisely the same black colour as those found in the peaty shell-marl of Scotland; and we might imagine them to have been dyed black in *Miocene peat* which was swept down into the sea during the waste of cliffs, did we not find the remains of cetacea in the same strata, bones, for example, of the lamantine, morse, sea-calf, and dolphin, having precisely the same colour."

[91] Lyell, *Principles of Geology*, vol. 1, p. 316: "We have before mentioned the violent earthquakes which, in 1812, convulsed the valley of the Mississippi at New Madrid, for the space of three hundred miles in length. As this happened exactly at the same time as the great earthquake of Caraccas, it is probably that these two points are parts of one continuous volcanic region...."

[92] Humboldt, *Personal Narrative*, vol. 4, pp. 11–12: "The extraordinary commotions felt almost continually during two years on the borders of the Missisippi and the Ohio, and which coincided in 1812 with those of the valley of Caraccas, were preceded at Louisiana by a year almost exempt from thunder storms."

[93] Lyell, *Principles of Geology*, vol. 1, pp. 321–322: "Syria and Palestine abound in volcanic appearances, and very extensive areas have been shaken, at different periods, with great destruction of cities and loss of lives. It has been remarked...that from the commencement of the thirteenth to the latter half of the seventeenth century, there was an almost entire cessation of earthquakes in Syria and Judea; and, during this interval of quiescence, the Archipelago, together with part of the adjacent coast of Lesser Asia, as also Southern Italy and Sicily, suffered extraordinary convulsions; while volcanic eruptions in those parts were unusually frequent. A more extended comparison...seems to confirm the opinion, that a violent crisis of commotion never

visits both at the same time. It is impossible for us to declare, as yet, whether this phenomenon is constant in this, or general in other regions, because we can rarely trace back a connected series of events farther than a few centuries; but it is well known that, where numerous vents are clustered together within a small area, as in the many archipelagos for instance, two of them are never in violent eruption at once."

[94] Jean Baptiste Bory de Saint-Vincent, *Voyage dans les quatre principales îles des mers d'Afrique...1801–1802* (Paris, 1804), vol. 1, chap. 6 describes the physical geography of Mauritius but does not answer Darwin's question directly. While at Mauritius Darwin was unable to inspect the entire island himself and sought information from other sources. See *VI*, pp. 28–31 and pp. 118–120 of this notebook.

[95] Alexander von Humboldt, *Fragmens de géologie et de climatologie asiatiques*. 2 vols. (Paris, 1831).

[96] The exact quotation is uncertain, but the following sentence suggests Humboldt's views (*Fragmens...asiatiques*, vol. 1, pp. 5–6): "La *volcanicité*, c'est-à-dire, l'influence qu'exerce l'intérieur d'une planète sur son enveloppe extérieure dans les différens stades de son refroidissement, à cause de l'inégalité d'agrégation (de fluidité et de solidité), dans laquelle se trouvent les matières qui la composent, cette action du dedans en dehors (si je puis m'exprimer ainsi) est aujourd'hui très affaiblie, restreinte à un petit nombre de points, intermittente, moins souvent déplacée, très simplifiée dans ses effets chimiques, ne produisant des roches qu'autour de petites ouvertures circulaires ou sur des crevasses longitudinales de peu d'étendue, ne manifestant sa puissance, à de grandes distances, que dynamiquement en ébranlant la croûte de notre planète dans des directions linéaires, ou dans des étendues (cercles d'oscillations simultanées) qui restent les mêmes pendant un grand nombre de siècles."

[97] This entry is written in brown ink.

[98] George Juan and Antonio de Ulloa, *A Voyage to South America* (4th ed.; London, 1806), vol. 2, p. 84: "According to an account sent to Lima after this accident, a volcano in Lucanas burst forth the same night and ejected such quantities of water, that the whole country was overflowed; and in the mountain near Patas, called Conversiones de Caxamarquilla, three other volcanoes burst, discharging frightful torrents of water...."

[99] George Poulett Scrope, *Considerations on Volcanos* (London, 1825), chap. 2, sections 41–42 including the statement on p. 60: "It is obvious how the powerful ascending draught of air which constitutes a hurricane, and which acts so strongly in depressing the barometer, will have an equal effect in setting loose the imprisoned winds of the earth." Also see *JR*, p. 431.

[100] Lyell, *Principles of Geology*, vol. 3, p. 364: "If...we conceive it probable that plutonic rocks have originated in the nether parts of the earth's crust, as often as the

volcanic have been generated at the surface, we may imagine that no small quantity of the former class has been forming in the recent epoch, since we suppose that about 2000 volcanic eruptions may occur in the course of every century, either above the waters of the sea or beneath them."

[101] John Michell, 'Conjectures concerning the Cause, and Observations on the Phænomena of Earthquakes; particularly of that Great Earthquake of the First of November 1755, which proved so fatal to the City of Lisbon, and whose Effects were felt as far as Africa, and more or less throughout almost all Europe', *Philosophical Transactions of the Royal Society of London*, vol. 51 (1760), p. 617: "The great earthquake that destroyed Lima and Callao in 1746, seems also to have come from the sea; for several of the ports upon the coast were overwhelmed by a great wave, which did not arrive till four or five minutes after the earthquake began, and which was preceded by a retreat of the waters, as well as that at Lisbon." Darwin's own copy of this article was a reprint which had been repaginated by the printer and is separately bound; this quotation appears on p. 54 of his copy.

[102] Lyell, *Principles of Geology*, vol. 1, pp. 471–472: "Sometimes the rising of the coast must give rise to the retreat of the sea, and the subsequent wave may be occasioned by the subsiding of the shore to its former level; but this will not always account for the phenomena. During the Lisbon earthquake, for example, the retreat preceded the wave not only on the coast of Portugal, but also at the island of Madeira and several other places."

[103] Lyell did discuss 'partial shrinking after elevation', but, as Darwin's cancellation indicates, did not relate it to the existence of an underlying injected mass of fluid rock. See Lyell, *Principles of Geology*, vol. 1, p. 477: "It is to be expected, on mechanical principles, that the constant subtraction of matter from the interior will cause vacuities, so that the surface undermined will fall in during convulsions which shake the earth's crust even to great depths, and the sinking down will be occasioned partly by the hollows left when portions of the solid crust are heaved up, and partly when they are undermined by the subtraction of lava and the ingredients of decomposed rocks." In his own copy of this work Darwin commented: "if there are hollows left what forces up the lava" and then crossed out his remark. A few pages previously (p. 468) he had challenged Lyell's association of the occurrence of submarine earthquakes with the percolation of sea water to underlying masses of incandescent lava with the remark, "We may more easily imagine the fluid stone injected (as occurs in every mountain chain) amongst damp strata." He also questioned whether water could percolate through strata already under great pressure. In short, it would seem that Darwin realized he was describing his own idea rather than Lyell's in the course of writing this entry.

[104] Lyell, *Principles of Geology*, vol. 1, p. 191: "Darby mentions beds of marine shells on the banks of Red River, which seem to indicate that Lower Louisiana is of

recent formation: its elevation, perhaps, above the sea, may have been due to the same series of earthquakes which continues to agitate equatorial America." The work referred to is William Darby, *A Geographical Description of the State of Louisiana* (Philadelphia, 1816).

[105] William J. Burchell, *Travels in the Interior of Southern Africa* (London, 1824), vol. 2, pp. 71–79 describes the killing of two rhinoceroses south of the Hyena Mountains (30° 10′ S., 24° 0′ E.). In his own copy of the work Darwin scored the passage on p. 78 where Burchell described his sensation of the heat on a day of the hunt: "Although so chilling at sunrise, the weather had, by noon, changed to the opposite extreme. Exposed in the middle of a dry plain, where not a tree to afford shade was to be seen, I scarcely could endure the rays of the sun, which poured down, as it were, a shower of fire upon us." See also *JR*, pp. 101–102.

[106] Lyell, *Principles of Geology*, vol. 3, p. 355: "The main body of the granite here [in Cornwall] is of a porphyritic appearance with large crystals of felspar; but in the veins it is fine-grained and without these large crystals.... The vein-granite of Cornwall very generally assumes a finer grain, and frequently undergoes a change in mineral composition, as is very commonly observed in other countries. Thus, according to Professor Sedgwick, the main body of the Cornish granite is an aggregate of mica, quartz, and felspar; but the veins are sometimes without mica, being a granular aggregate of quartz and felspar."

[107] Daubeny, *Volcanos*, pp. 94–95: "Trachytic porphyry also appears to pass by imperceptible gradations into the next species, pearlstone, which is characterized by the vitreous aspect generally belonging to its component parts.... In its simplest form, this rock presents an assemblage of globules, varying from the size of a nut to that of a grain of sand, which have usually a pearly lustre, and scaly aspect.... In some varieties the globules are destitute of lustre, and exhibit at the same time sundry alterations in their size, structure, and mode of aggregation, till at length they entirely disappear, and the whole mass puts on a stony appearance, which retains none of the characters of pearlstone.... Various alternations occur between the glassy and stony varieties of the pearlstone, sometimes so frequent as to give a veined or ribboned appearance to the rock, at others curiously contorted as though they had been disturbed in the act of cooling."

[108] William Phillips, *An Elementary Introduction to the Knowledge of Mineralogy* (3rd ed.; London, 1823) contains no reference to pearlstone in Peru, but on p. 112 there is the statement that "At Tokay in Hungary, [pearlstone] is found enclosing round masses of black vitreous obsidian, and is intermixed with the debris of granite, gneiss, and porphyry, and alternating in beds with the latter."

[109] Daubeny, *Volcanos*, p. 180: "[The island of Ischia] is composed for the most part of a rock which seems to consist of very finely comminuted pumice, reagglutinated

so as to form a tuff. ...Although the pumiceous conglomerate, as I shall venture to call this rock, is seen in every part of the island, yet at Monte Vico...we observe intermixed with it huge blocks of trachyte...." In a footnote on this page, which Darwin heavily scored in his own copy of the work, Daubeny stated that other geologists had identified the predominate rock at Ischia as an "earthy variety of trachyte".

[110] Sir George Steuart Mackenzie, F.R.S. (1780–1848), mineralogist, as quoted in Daubeny, *Volcanos*, p. 221: "In many places [in Iceland], [Sir G. Mackenzie] says, an extensive stratum of volcanic matter has been heaved up into large bubbles or blisters, varying from a few feet to forty or fifty in diameter." The original reference is to George Steuart Mackenzie, *Travels in the Island of Iceland...1810* (Edinburgh 1811), pp. 389–390.

[111] As quoted in Daubeny, *Volcanos*, p. 313: "In Sumatra, Marsden has described four [volcanos] as existing, but the following are all the particulars known concerning them: Lava has been seen to flow from a considerable volcano near *Priamang*, but the only volcano this observer had an opportunity of visiting, opened on the side of a mountain about 20 miles inland of Bencoolen, one fourth way from the top, so far as he could judge....He never observed any connexion between the state of the mountain and the earthquake, but it was stated to him, that a few years before his arrival it was remarked to send forth flame during an earthquake, which it does not usually do. The inhabitants are however alarmed, when these vents all remain tranquil for a considerable time together, as they find by experience, that they then become more liable to earthquakes." The original reference is to William Marsden, *The History of Sumatra* (3rd ed.; London, 1811), pp. 29–30.

[112] Alexandre Moreau de Jonnès (1778–1870), French economist and natural historian as cited in Daubeny, *Volcanos*, p. 334: "The process, by which these islands, according to Moreau de Jonnes, are in many instances formed, is sufficiently curious; first a submarine eruption raises from the bottom of the sea masses of volcanic products, which, as they do not rise above the surface of the water, but form a shoal a short way below its surface, serve as a foundation on which the Madreporites and other marine animals can commence their superstructure. Hence those beds of recent coralline limestone, seen covering the volcanic matter in many of the islands." The original reference in this case is to Alexander von Humboldt who had communicated directly with Moreau de Jonnès on the subject. See Humboldt, *Personal Narrative*, vol. 4, pp. 42–43; also M. Cortès and Alexandre Moreau de Jonnès, 'Mémoire sur la géologie des Antilles', *Journal de physique, de chimie, d'histoire naturelle et des arts*, vol. 70 (1810), pp. 130–131.

[113] Roussin, *Le Pilote du Brésil*, p. 47 states that on approaching the banks of Cape S. Roque: "...nous croyons avoir observé que le sable est d'autant plus rare et les graviers d'autant plus communs, que les sondes sont plus petites et plus voisines des bancs."

[114] Humphry Davy, 'On the corrosion of copper sheeting by sea water, and on methods of preventing this effect; and on their application to ships of war and other ships', *Philosophical Transactions of the Royal Society of London*, vol. 114 (1824), pp. 151–158. After describing his experiments Davy concluded on p. 158: that "small quantities of zinc, or which is much cheaper, of malleable, or cast iron, placed in contact with the copper sheeting of ships, which is all in electrical connection, will entirely prevent its corrosion. And as negative electricity cannot be supposed favourable to animal or vegetable life; and as it occasions the deposition of magnesia, a substance exceedingly noxious to land vegetables, upon the copper surface; and as it must assist in preserving its polish, there is considerable ground for hoping that the same application will keep the bottoms of ships clean, a circumstance of great importance both in trade and naval war."

[115] See note 59.

[116] The entries pertaining to Fig. 5 are written in brown ink.

[117] A bar and a dot over a number indicates that no bottom was found at that depth. All entries on this page are in brown ink, except for the page number.

[118] Thomas Sorrell (c. 1797–?), boatswain of the H.M.S. *Beagle*; personal communication. See Fitzroy, ed., *Narrative of the Surveying Voyages of His Majesty's Ships Adventure and Beagle*, vol. 2, p. 21. Also see *JR*, p. 282.

[119] This entry is written in light brown ink.

[120] Humboldt, *Personal Narrative*, vol. 4, p. 384: "We discover between Calabozo, Uritucu, and the *Mesa de Pavones*, wherever men have made excavations of some feet deep, the geological constitution of the Llanos. A formation of red sandstone [*Rothes todtes liegende*] (or ancient conglomerate) covers an extent of several thousand square leagues. We shall find it again hereafter in the vast plains of the Amazon, on the eastern boundary of the province of Jaën de Bracamoros. This prodigious extension of red sandstone, in the low grounds that stretch along the East of the Andes, is one of the most striking phenomena, with which the study of rocks in the equinoctial regions furnished me."

[121] Miers, *Travels in Chile and La Plata*, vol. 1, pp. 394–395: "All around Quintero [near Quillota]. . . the fishermen had employed themselves digging shells for lime-making from a stratum four or five feet thick, in the recesses of the rocks, at the height of fifteen feet above the usual level of the sea, it being evident that at no very distant period this spot must have been buried in the sea, and uplifted probably by convulsions similar to the one now described." Also p. 458: "The recent shelly deposites mixed with loam [at Quintero] I have traced to places three leagues from the coast, at a height of 500 feet above the level of the sea. . . ." See *GSA*, p. 35.

[122] Lyell, *Principles of Geology*, vol. 3, p. 371: "[According to M. Elie de Beaumont]...near Champoleon [in France], a granite composed of quartz, black mica, and rose-coloured felspar, is observed partly to overlie the secondary rocks, producing an alteration which extends for about thirty feet downwards, diminishing in the inferior beds which lie farthest from the granite....In the altered mass the argillaceous beds are hardened, the limestone is saccharoid, the grits quartzose, and in the midst of them is a thin layer of an imperfect granite. It is also an important circumstance, that near the point of contact both the granite and the secondary rocks become metalliferous, and contain nests and small veins of blende, galena, iron, and copper pyrites."

[123] Fitton, 'Geology' as quoted in note 47.

[124] Lesson and Garnot, *Voyage autour du monde...Zoologie*, vol. 1, part 1, p. 5: "Toutes les côtes de la Nouvelle-Galles du Sud [New South Wales] sont, en effet, entièrement composées d'un grès houiller à molécules peu adhérentes; et ce que nous appelons le premier plan des montagnes Bleues est également composé de ce grès, qui cesse entièrement au mont York. Là, une vallée profonde isole ce premier plan du second, qui est composé en entier de granite."

[125] Lesson and Garnot, *Voyage autour du monde... Zoologie*, vol. 1, part 1 [1826] and part 2 [1828]. The following citations pertain to the entire paragraph on p. 102 of the notebook:

On the formations of Payta see part 1, pp. 260–261: "Le lambeau de sol tertiaire se compose de couches ou bancs alternatifs, dont voici l'énumération, en commençant par la formation de phyllade qui le supporte. 1° *Roches talqueuses phylladiformes*, terrain primordial. 2° *Argiles plastiques.* — Sable argileux, schisteux, traversé par des veines entrecroisées de gypse fibreux...3° *Calcaire grossier*..." Rock cleavage is described as running from east to west on p. 260. On p. 262 Lesson uses the figure 200 feet in describing the change in sea level which would have caused such configurations of strata as seen at Payta.

With respect to volcanic formations on the north part of New Zealand, there is Lesson's remark in part 2, p. 410 that "De nombreux volcans, dont les traces des éruptions sont récentes, existent sur plusieurs points de ces îles [off the north shore of the North Island]...Aussi trouve-t-on communément des pierres ponces...." With respect to richness of plant genera in New Zealand see the quotation from Lesson in note 80.

On St Catherine's see part 1, p. 189: "Le granite forme entièrement la croûte minérale de l'île de Sainte-Catherine et du continent voisin...."

On the Falkland Islands see part 1, p. 198–199: "Les couches se composent de feuillets fendillés dans tous les sens, dont la direction, au lieu d'être horizontale, est presque verticale, et forme particulièrement sur le pourtour de la baie un angle de 45 degrés: ceux de la grande terre se dirigent à l'Est, et ceux des îlots aux pingoins à l'Ouest. . . . Cette phyllade supporte un grès schisteux. . . ." Also on p. 200 reference is made to the discussion by "Pernetty" of a "montagne des Ruines" which looked man-made, and on p. 201, to what Darwin later quoting directly from Pernety called a "stream of stones" and what Lesson referred to as "blocs énormes du même grès, entassés pêle-mêle. . ." See *JR*, p. 255 and Antoine Joseph Pernety, *Journal historique d'un voyage. . . aux îles Malouines* (Berlin, 1769), vol. 2, p. 526.

On the region around Concepcion see part 1, p. 231: "La couche la plus inférieure est formée par une sorte de phyllade noire, compacte et terne; celle qui est moyenne se compose d'un mica-schiste à feuillets très-brillants, dont la direction est de l'Ouest à l'Est." The presence of talcose slates at Concepcion is mentioned on p. 232.

[126] Juan and Ulloa, *A Voyage to South America*, vol. 2, p. 97.

[127] Juan and Ulloa, *A Voyage to South America*, vol. 2, p. 147: "These are the principal mines of Potosi, but there are several smaller crossing the mountain on all sides. The situation of the former of these mines is on the north side of the mountain, their direction being to the south, a little inclining to the west; and it is the opinion of the most intelligent miners in this country, that those which run in these directions are the richest."

[128] Juan and Ulloa, *A Voyage to South America*, vol. 2, p. 252: "The country round the bay, particularly that between Talcaguana and Conception. . . is noted for . . . a stratum of shells of different kinds, two or three toises in thickness, and in some places even more, without any intermixture of earth, one large shell being joined together by smaller, and which also fill the cavities of the larger. . . . Quarries of the same kind of shells, are found on the tops of mountains in this country, fifty toises above the level of the sea." Also, p. 254: "All these species of shellfish are found at the bottom of the sea in four, six, ten and twelve fathom water. They are caught by drags; and. . . no shells, either the same, or that have any resemblance to them, are seen either on the shores continually washed by the sea, or on those tracks which have been overflowed by an extraordinary tide."

[129] John Playfair, *Illustrations of the Huttonian Theory of the Earth* (Edinburgh, 1802), pp. 51–52: "Indeed, the interposition of a breccia between the primary and secondary strata, in which the fragments, whether round or angular, are always of the primary rock, is a fact so general, and the quantity of this breccia is often so great, that it leads to a conclusion more paradoxical than any of the preceding, but from which, nevertheless, it seems very difficult to with-hold assent. Round gravel, when in great abundance, agreeably to a remark already made, must necessarily be considered as a production peculiar to the beds of rivers, or the shores of continents, and as hardly ever formed at great depths under the surface of the sea. It should seem, then, that under the primary schistus, after attaining its erect position, had been raised up

to the surface, where this gravel was formed; and from thence had been let down again to the depths of the ocean, where the secondary strata were deposited on it. Such alternate elevations and depressions of the bottom of the sea, however extraordinary they may seem, will appear to make a part of the system of the mineral kingdom, from other phenomena hereafter to be described."

[130] The principle expressed in this passage, that the destruction of the earth's surface is required for its renovation, is consistent with the general content of the work of the great British geologist and member of the Royal Society of Edinburgh, James Hutton (1726–1797). However, as Darwin's cancellation would seem to indicate, the application of the principle failed in this instance, for Hutton, speaking providentially, had chosen rather to characterize volcanos as instruments designed "to prevent the unnecessary elevation of land, and the fatal effects of earthquakes", and his interpreter John Playfair, F.R.S. (1748–1819), while not quoting Hutton's words, did not challenge his conclusion. See James Hutton, *Theory of the Earth* (Edinburgh, 1795), vol. 1, p. 146, and Playfair, *Illustrations of the Huttonian Theory*, pp. 116–119. Later Charles Lyell (note 3) was more sanguine on the subject of the fatal effects of earthquakes. See his *Principles of Geology*, vol. 1, p. 479.

[131] Sir Richard Owen, F.R.S. (1804–1892), comparative anatomist and palaeontologist. In 1837 Owen was Assistant Conservator in the Hunterian Museum of the Royal College of Surgeons, and first Hunterian Professor of Comparative Anatomy and Physiology at the Royal College of Surgeons. In 1856 Owen left the Royal College of Surgeons for the British Museum, where he served as Superintendent of the Natural History Departments of the Museum and then later, as Superintendent of the new Natural History Museum in South Kensington. Owen's palaeontological work began in 1837 with his studies of Darwin's collection of South American fossil mammals. For Darwin's account of the opening of negotiations with Owen with respect to collections from the *Beagle* voyage see the letter from Darwin to J. S. Henslow, dated 3 October 1836 in Nora Barlow, ed., *Darwin and Henslow: The Growth of an Idea* (Berkeley and Los Angeles, 1967), pp. 118–119. Owen's completed work on Darwin's specimens is contained in Richard Owen, *The Zoology of the Voyage of H.M.S. Beagle . . . 1832–1836 . . . Edited and Superintended by Charles Darwin. Part I: Fossil Mammalia.* 4 numbers. (London, 1838–1840).

[132] [Capt.] John Ross, *A Voyage of Discovery. . . for the Purpose of Exploring Baffin's Bay* (London, 1819), p. 178: "Soundings were obtained correctly in one thousand fathoms [at Possession Bay], consisting of soft mud, in which there were worms. . . . The temperature of the water on the surface was 34 1/2° [F.], and at eighty fathoms 32°; . . . at two hundred and fifty fathoms [measurement taken aboard another ship], . . . 29 1/2° [F.]." In Appendix No. III, p. lxxxv this information is summarized and the coordinates of Possession Bay given as 73° 39′ N., 77° 08′ W.

[133] These observations by John Herschel (note 40) on the subject of the crystallization of barium sulphate were probably communicated to Darwin by Charles Lyell (note 3). Herschel's letter of 20 February 1836 to Lyell, quoted in note 40, contains the following passage (Cannon, 'The Impact of Uniformitarianism', p. 310): "Cleavages of Rocks.—If Rocks have been heated to a point admitting a commencement of crystallization, ie to the point where particles can begin to move inter se—or at least on their own axes—some general cause must determine the position these particles will rest in on cooling—probably position will have some relation to the direction in which the heat escapes.—Now when all—or a majority of particles of the same nature have a general tendency to one position that must of course determine a cleavage plane.—Did you never notice how the infinitesimal crystals of fresh precipitated sulphate of Baryta [barium sulphate] & some other such bodies—arrange themselves alike in the fluid in which they float so as, when stirred all to glance with one light & give the appearance of *silky filaments*. Ask Faraday to shew you this phenomenon if you have not seen it—it is very pretty. What occurs in our expt, on a minute scale may occur in nature on a great one, as in granites, gneisses, mica slates &c—some sorts of soap in which insoluble margarates exist shew it beautifully [added: when mixed with water]." Lyell incorporated Herschel's observation into his next edition of the *Principles*. See *Principles of Geology* (5th ed.; London, 1837), vol. 4, pp. 358. Presumably Lyell showed Darwin Herschel's letter, or discussed its contents with him, sometime in late 1836 or early 1837. The 'Faraday' referred to in Herschel's letter is Michael Faraday, F.R.S. (1791–1867), the eminent natural philosopher and experimentalist.

[134] Erasmus Alvey Darwin (1804–1881). Charles' older brother who pursued the study of chemistry in his youth and early manhood.

[135] José de Acosta, *Histoire naturelle et moralle des Indes* (Paris, 1600), p. 125 refers to "des tremblemens de terre qui ont couru depuis Chillé, jusques à Quitto, qui sont plus de cinq cens lieues..." Acosta continued, "En la coste de Chillé (il ne me souvient quelle année) fut un tremblement de terre si terrible....A peu de temps delà, qui fut l'an, de quatre vingts deux, vint le tremblement d'Arequipa, qui abbatit & ruina presque toute cette ville là. Du depuis en l'an quatre vingts six...aduint un autre tremblement en la cité des Roys [Lima]...." And on p. 125 verso: "En apres l'an enfuyuant, il y eut encor un autre tremblement de terre au Royaume & cité de Quitto, & semble que tous ces notables tremblemens de terre en ceste coste, ayent succedé les uns aux autres par ordre...."

[136] Joseph-Charles Bailly (1777–1844), mineralogist to the expedition, as quoted in François Péron, *Voyage de découvertes aux terres australes...1800–1804* (Paris, 1807), vol. 1, pp. 54–55. Following the passage quoted which Darwin copied correctly except for one misspelling ("d'lile" for "de l'île") and the loss of a few accent marks, the text continues (p. 55): "De ces observations, il résulte bien incontestablement que

toutes ont la même origine, qu'elles datent toutes de la même époque; que réunies jadis, elles n'ont pu être séparées depuis, que par quelque révolution violente et subite. Quelle peut avoir été cette dernière révolution?...Tous les fait se réunissent pour prouver que l'île toute entière ne formoit jadis qu'une énorme montagne brûlante; qu'épuisée, pour ainsi-dire, par ses éruptions, elle s'affaissa sur elle-même, engloutit dans ses abîmes la plus grande partie de sa propre masse, et que de cette voûte immense, il ne resta debout que les fondemens, dont les débris entr'ouverts sur différens points, forment les montagnes actuelles de l'île. Quelques pitons de forme conique, qui s'élèvant vers le centre du pays, notamment le Piton du centre, portent les caractères d'une origine postérieure à l'éboulement du cratère...." Also see *VI*, pp. 29–31.

[137] Bailly (note 136) as quoted in Péron, *Voyage de découvertes aux terres australes*, vol. 1, p. 295: "De hautes montagnes granitiques...dont les sommités étoient presque entièrement nues, forment toute la côte orientale de cette partie de la terre de Diémen...." Also see p. 304 for a description of more of the east coast of Van Diemen's Land [Tasmania].

[138] Henry Bolingbroke, *A Voyage to the Demerary* (London, 1807), p. 200 contains the passage Darwin quotes and pp. 200–201 the additional comment: "This constant shooting upwards of the land, which is so sensible in the West Indies, has been little heeded by European mineralogists."

[139] Webster, *Narrative of a Voyage to the Southern Atlantic Ocean*, vol. 1, p. 371: "Instances of earthquakes occuring in the island [St Helena] are on record. One took place in 1756, and in June 1780. On the 21st September 1817, one occurred, which it is said was particularly noticed by Napoleon, who thought that the Conqueror, 74, in which he had been, was blown up." The reference to antarctic vegetation pertains to Webster's discussion of the natural history of Cape Horn, Staten Island, and Deception Island in vol. 2, pp. 290–306.

[140] Darwin apparently searched Juan and Ulloa's *A Voyage to South America* for evidence connecting Indian habitation and climatic change, and could not find it. He was more successful in his reading of Antonio de Ulloa's *Noticias americanas* (2nd ed.; Madrid, 1792). He later quoted from that work (p. 302) in translation, presumably his own, to the effect that Indians of one arid region in the Andes had lost the art of making durable bricks from mud. This suggested to Darwin that the local climate had once been wetter, which fitted his notion that the South American continent had undergone elevation in geologically recent times. See *JR*, pp. 409–411.

[141] Edmond Temple, *Travels in Various Parts of Peru, Including a Year's Residence in Potosi* (London, 1830), vol. 2, p. 10: "In the course of this day's journey were to be seen, in well-chosen spots, many Indian villages and detached dwellings, for the most part in ruins. Up even to the very tops of the mountains, that line the

valleys through which I have passed, I observed many ancient ruins, attesting a former population where now all is desolate." For passages on a similar theme see pp. 4 and 5. Also consult *JR*, p. 412, where the quotation from Temple appears in slightly different form. From his comment it would appear that Charles' sister Caroline Darwin Wedgwood (1800–1888) gave him the reference to Temple.

[142] John W. Webster, *A Description of the Island of St. Michael* (Boston, 1821), p. 124: "There is scarcely a man on the island, who has not a dog, and many have half a dozen. It is a remarkable fact that, although these animals are so numerous, no instance of hydrophobia was ever known among them." See *JR*, p. 436.

[143] Sir Woodbine Parish, F.R.S. (1796–1882), personal communication. See *JR*, p. 156: "Sir Woodbine Parish informed me of another and very curious source of dispute [in the province of Buenos Ayres]; the ground being so long dry, such quantities of dust were blown about, that in this open country the landmarks become obliterated, and people could not tell the limits of their estates." Parish served as commissioner and consul general and then chargé d'affaires to Buenos Ayres from 1823–1832. Upon returning to London he became active in scientific societies. He was a long-time vice-president of the Royal Geographical Society and served on the Council of the Geological Society of London from 1834–1841, being sometime vice-president and during 1835–1836 one of the secretaries.

[144] Félix d'Azara, *Voyages dans l'Amérique Méridionale...1781–1801* (Paris, 1809), vol. 1, p. 374: "On voit un exemple aussi étonnant de cette fougue dans les années sèches, où l'eau est extrêmement rare au sud de Buenos-Ayres. En effet, ils partent comme fous, tous tant qu'ils sont, pour aller chercher quelque mare ou quelque lac: ils s'enfoncent dans la vase, et les premiers arrivés sont foulés et écrasés par ceux qui les suivent. Il m'est arrivé plus d'une fois de trouver plus de mille cadavres de chevaux sauvages morts de cette façon." See *JR*, p. 156.

[145] John Hunter, *An Historical Journal of the Transactions at Port Jackson and Norfolk Island...since the publication of Phillip's Voyage* (London, 1793), pp. 507, 508, 525, and 535 refer to the drought around Sydney in the first half of the year 1791. 'Phillip's Voyage' refers to the account of his travels written by Arthur Phillip (1738–1814), vice-admiral and first governor of New South Wales, published as *The Voyage of Governor Phillip to Botany Bay* (London, 1789).

[146] Charles Sturt, *Two Expeditions into the Interior of Southern Australia... 1828–1831* (London, 1833), vol. 1, p. 1: "The year 1826 was remarkable for the commencement of one of those fearful droughts to which we have reason to believe the climate at New South Wales is periodically subject. It continued during the two following years with unabated severity." And p. 2: "But, however severe for the colony the seasons had proved...it was borne in mind at this critical moment, that the wet and swampy state of the interior had alone prevented Mr. Oxley from

penetrating further into it, in 1818.... As I had early taken a great interest in the geography of New South Wales, the Governor was pleased to appoint me to the command of this expedition." See also *JR*, p. 157.

147 From this entry it would appear likely that it was Richard Owen (note 131) who referred Darwin to the article by Thomas Rackett, 'Observations on *Cancer salinus*', which is quoted enthusiastically in *JR*, p. 77: "In the Linnean [Society of London] Transactions, [1815], vol. xi, p. 205, a minute crustaceous animal is described, under the name of *Cancer salinus*. It is said to occur in countless numbers in the brine pans at Lymington; but only in those in which the fluid has attained, from evaporation, considerable strength; namely about a quarter of a pound of salt to a pint of water. This cancer is said, also, to inhabit the salt lakes of Siberia. Well may we affirm, that every part of the world is habitable!"

148 Rev. John Stevens Henslow (1796–1861), Professor of Botany at Cambridge University and Darwin's 'Master in Natural History'. (Letter from Darwin to Henslow, January 1836, in Nora Barlow, ed., *Darwin and Henslow*, p. 114.) Henslow himself did not publish on the subject of springs, but he may have been the source for two references which Darwin quoted on the subject in the *JR*, p. 78. Both works cited discuss plant life at the location of the springs, a subject which would have interested Henslow. The references were to James Edward Alexander, 'Notice regarding the Salt Lake Inder, in Asiatic Russia', *Edinburgh New Philosophical Journal*, vol. 8 (1830), pp. 18–20; and Peter Simon Pallas, *Travels through the Southern Provinces of the Russian Empire...1793–1794* (London, 1802), vol. 1, pp. 129–134. The entry 'Springs. (Henslow)' is written in brown ink, the preceding entry on 'Mr Owen' in pencil.

149 The two ostriches are the greater or common rhea, *Rhea americana*, found from north-eastern Brazil to the Rio Negro in central Argentina, and the lesser rhea or Darwin's rhea, *Pterocnemia pennata*, found in the Patagonian lowlands, where Darwin collected portions of a specimen, and in the high Andes of Peru, Bolivia, northern Chile, and northwestern Argentina. The lesser rhea became known as Darwin's rhea following its identification by the ornithologist John Gould, F.R.S. (1804–1881) at a meeting of the Zoological Society of London on 14 March 1837. Gould was then unaware that the species had already been described in 1834 by the French naturalist Alcide Dessalines d'Orbigny (1802–1857). For Gould's report on *Rhea darwinii* and comments by Darwin on the habits of the two species (but primarily the common rhea) and on their geographical distribution see the *Proceedings of the Zoological Society of London*, vol. 5 (1837), pp. 35–36. For further treatment see John Gould, *The Zoology of the Voyage of H.M.S. Beagle...1832–1836...Edited and Superintended by Charles Darwin. Part III: Birds.* 5 numbers. (London, 1838–1841), pp. 120–125 including plate. Also see *JR*, pp. 108–110.

Rhea americana, the bigger or common 'ostrich' referred to on pages 127 and 130.

Darwin's rhea, *Pterocnemia pennata*, the smaller 'ostrich' or 'Petisse' referred to on pages 127 and 130. Of this rhea Darwin wrote: 'This species... differs in many respects from the *Rhea Americana*. It is smaller, and the general tinge of the plumage is a light brown in place of grey; each feather being conspicuously tipped with white. The bill is considerably smaller, and especially less broad at its base; the culmen is less than half as wide, and becomes slightly broader towards the apex, whereas in the *R. Americana* it becomes slightly narrower; the extremity, however, of both the upper and lower mandible, is more tumid in the latter, than in the *R. Darwinii*.... The skin round and in front of the eyes is less bare in *R. Darwinii*; and small bristly feathers, directed forwards, reach over the nostrils. The feet and tarsi are nearly of the same size in the two species. In the *R. Darwinii*, short plumose feathers extend downwards in a point on the sides of the tarsus, for about half its length. The upper two-thirds of the tarsus, in front, is covered with reticulated scales in place of the broad transverse band-like scales of the *R. Americana*; and the scales of the lower third are not so large as in the latter. In the *R. Darwinii* the entire length of the back of the tarsus is covered with reticulated scales, which increase in size from the heel upwards: in the common *Rhea*, the scales on the hinder side of the tarsus are reticulated only on the heel, and about an inch above it; all the upper part consisting of transverse bands, similar to those in front.' Quoted from John Gould, *The Zoology of the Voyage of H.M.S. Beagle. Part III: Birds*. 5 numbers. (London, 1838–1841), pp. 123–124.

Rhea Darwinii [*Pterocnemia pennata*], Plate 47 from John Gould, *The Zoology of the Voyage of H.M.S. Beagle. Part III: Birds*. 5 numbers. (London, 1838–1841). Drawing by John Gould, lithograph by Elizabeth Coxen Gould.

[150] One of the sources which Darwin likely drew on for this passage was Webster, *Voyage to the Southern Atlantic*, vol. 2, pp. 281–302.

[151] *Zorrilla* is the Spanish word for skunk. The notes on species ranges of South American forms which Darwin suggested making in this entry are presumably those found in the Darwin MSS, Cambridge University Library, vol. 29 (i). The 'Birds' list is numbered fol. 41; the 'Animals' list appears between fols. 46–47. The *zorrilla* appears on the list for animals.

[152] The extinct llama is the *Macrauchenia patachonica* as described by Richard Owen (note 131). See Owen, *The Zoology of the Voyage of H.M.S. Beagle. Part I: Fossil Mammalia*, pp. 10–11, 35–56 and plates VI–XV. Darwin collected the fossil specimens in January 1834 at the port of San Julián, having "no idea at the time, to what kind of animal these remains belonged." (*JR*, p. 208.) Owen's earliest known comment on the specimens occurs in a letter to Charles Lyell dated 23 January 1837 where he described them as follows:

> RUMINANTIA
> Fam: *Camelidae*
> 2 cervical vertebrae, portions of femur, & fragments of a Gigantic Llama! as large as a Camel, but an *Auchenia* (from the plains of Patagonia)

For the citation from Owen see Leonard G. Wilson, *Charles Lyell: The Years to 1841* (New Haven and London, 1972), p. 437. Also see *JR*, pp. 208–209. Several of the fossilized bones which Darwin collected of *Macrauchenia patachonica* are presently on display in the Fossil Mammal Gallery of the British Museum (Natural History). A number of Darwin's fossil mammalia came to the British Museum (Natural History) during World War II after the specimens had suffered damage from bombs which fell on the museum of the Royal College of Surgeons, where the specimens had been stored since the time Owen first worked on them. Illustrations on pages 112–114.

[153] The 'Petisse' is the lesser rhea, or Darwin's rhea. (See note 149.) Darwin customarily referred to the two rheas in his field notes as 'Avestruz' and 'Avestruz Petise' from the Spanish *avestruz* (ostrich) and *avestruz petiso* (small ostrich).

Here and for the following notes (154, 156–159) I am indebted to Dr David Snow of the British Museum (Natural History) at Tring for supplying the present-day identifications of Darwin's specimens. Names follow or are consistent with the usage in Rodolphe Mayer De Schauensee, *The Species of Birds of South America and Their Distribution* (Narberth, Pennsylvania, 1966).

[154] *Fourmilier*, ('antbird') so named by the French naturalist George-Louis Leclerc, Comte de Buffon (1707–1788), for its falsely reported habit of living chiefly on ants (*fourmis*), is a general term for a member of the essentially tropical American family Formicariidae. Since Darwin did not collect extensively in tropical areas and does not seem to have used the term *Fourmilier* elsewhere in his notes, it is doubtful

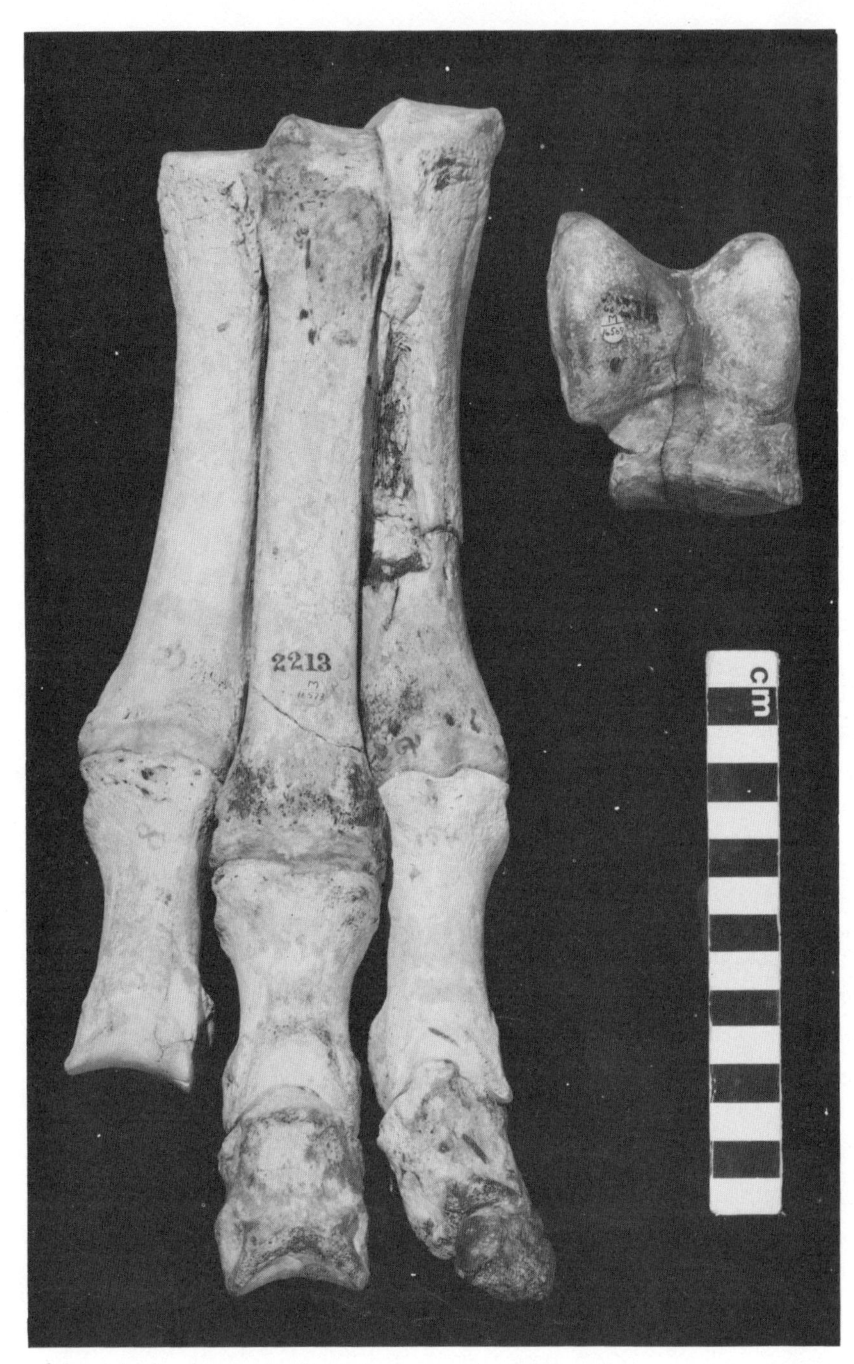

Bones of the right forefoot and ankle-joint (astragalus) of Darwin's specimen of *Macrauchenia patachonica* from the palaeontological collections of the British Museum (Natural History).

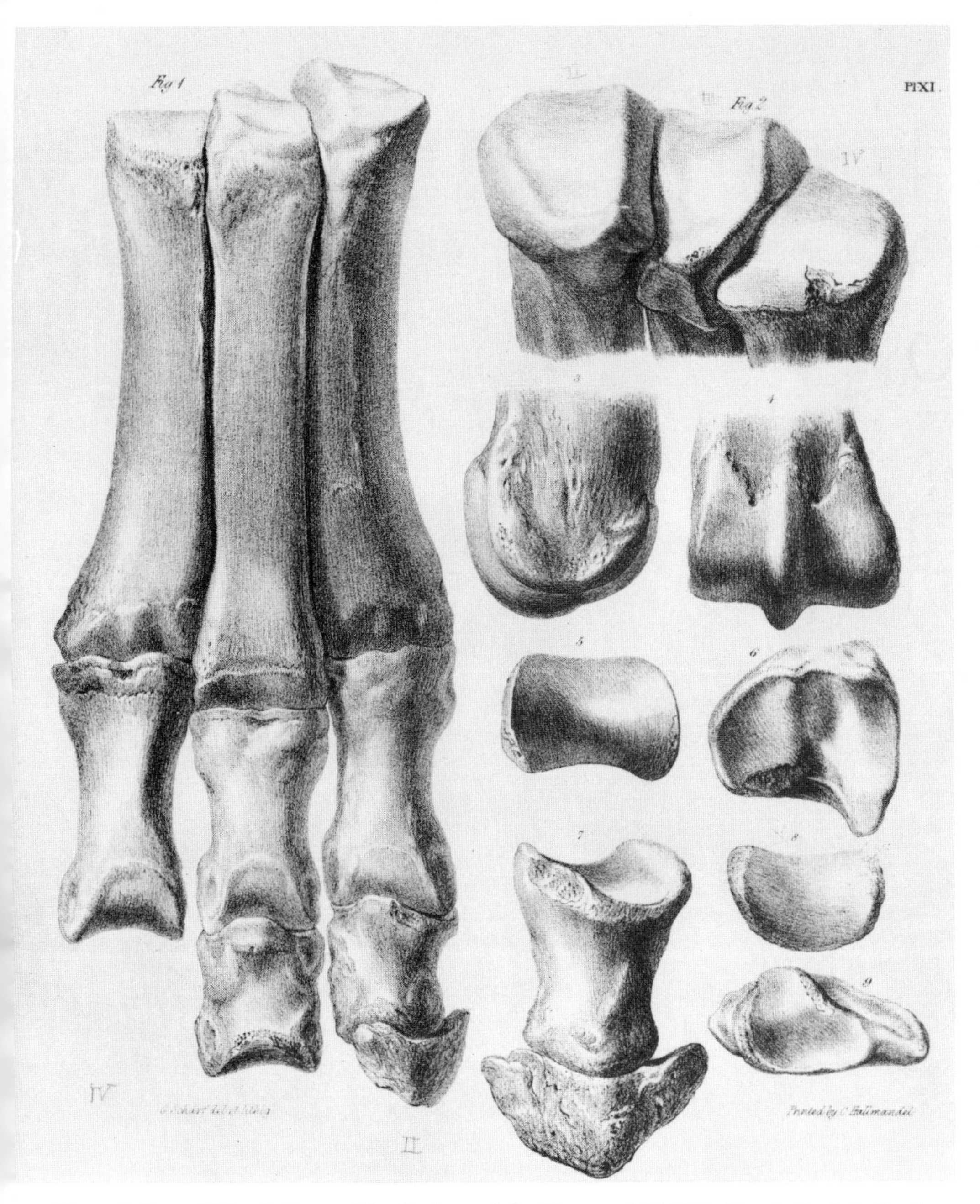

Plate 11 from Richard Owen, *The Zoology of the Voyage of H.M.S. Beagle. Part I: Fossil Mammalia*. 4 numbers. (London, 1838-1840). The plate contains figures of the bones of the right forefoot of Darwin's specimen of *Macrauchenia patachonica*. Drawing and lithograph by George Scharf.

A reconstruction of a specimen of the *Macrauchenia patachonica*, the 'extinct Llama' or 'extinct Guanaco' referred to on pages 129 and 130.

that it was the antbird, or at least primarily the antbirds, which he had in mind when he made this entry. More likely he was thinking of those birds which he described in his Ornithological Notes as *Myothera*, a term which was given as the equivalent of

Fourmilier in the '*Dict. classique*', the systemmatic work he had with him aboard ship. (See Jean Baptiste Bory de Saint-Vincent, ed., *Dictionnaire classique d'histoire naturelle* [Paris, 1825], vol. 7, pp. 22–25.) Under *Myothera* Darwin described a number of birds now assigned to the South American family Rhinocryptidae, a small family that appears to be closest to the Formicariidae, though its affinities are not certain. Chile is the centre of the present-day distribution of this family, in the sense that more genera occur there than in other countries. In general Darwin's recognition of specific differences and geographical ranges for this group was accurate and complete for the areas he visited. On Darwin's return to England the birds he had collected from this group were classified by John Gould (note 149) of the Zoological Society of London chiefly according to the taxonomy provided for the group by the German naturalist Friedrich Heinrich von Kittlitz (1799–1874) in 1830. For Darwin's discussion of the group, including Gould's classifications, see *JR*, pp. 329–330 and 351–353. For Darwin's listing of these specimens by number, all under the rubric *Myothera*, see Nora Barlow, ed., 'Darwin's Ornithological Notes', *Bulletin of the British Museum (Natural History)* Historical Series, vol. 2 (1963), pp. 201–278. Selected specimens which Darwin collected in Chile from this family, identified by their modern names, are listed as follows. Entries denoted with an asterisk indicate specimens collected by Darwin in the collections of the British Museum (Natural History) at Tring. The numbers given are those Darwin assigned to his specimens. From the genus *Pteroptochos* Darwin collected two species, the more southern *P. tarnii*, the Black-throated Huet-huet (specimen 2531*), and the more northern *P. megapodius*, the Moustached Turca (specimen 2172). [The Museum owns a specimen collected by Darwin of *P. megapodius* which no longer bears its original specimen number.] From the genus *Scelorchilus* Darwin collected three species, the more southern *S. rubecula*, the Chucao Tapaculo (specimen 2556*); the more northern *S. albicollis*, the White-throated Tapaculo (specimens 2173, 2174), and what was probably the northern desert subspecies *S.a. atacamae* (specimen 2825). Darwin also collected specimens of *Eugralla paradoxa*, the Ochre-flanked Tapaculo (specimen 2555*), whose range in Chile lies within that of *S. rubecula*. Further, he correctly identified the affinity of the 'black wren' of Tierra del Fuego (*Scytalopus fuscus* of Gould) with these other birds. This bird is now described as *S. magellanicus*, the Andean Tapaculo, and is assigned to the family Rhinocryptidae. In Chile it ranges from the Cape Horn Archipelago to Atacama. For general information on the family Rhinocryptidae, including drawings of various species and descriptions of their distinctive cries, see A. W. Johnson, *The Birds of Chile and Adjacent Areas of Argentina, Bolivia and Peru* (Buenos Aires, 1967), vol. 2, pp. 201–222. On these birds also see Gould, *The Zoology of the Voyage of H.M.S. Beagle. Part III: Birds*, pp. 70–74.

[155] The 'extinct Guanaco' is identical to the 'extinct Llama'. See note 152.

[156] The Chiloé creeper is *Aphrastura spinicauda*, the Thorn-tailed Rayadito (specimens 2129 and 2130). It is distributed from Coquimbo in Chile south to Tierra

del Fuego, and in Argentina from Neuquén and Rio Negro southwards. It also occurs on various off-lying islands including Chiloé. The Chiloé race is the distinctive subspecies *A. spinicauda fulva*, being buff-coloured instead of mainly white below. For further information on Darwin's specimens see 'Darwin's Ornithological Notes'; *JR*, p. 301; and Gould, *The Zoology of the Voyage of H.M.S. Beagle. Part III: Birds*, p. 81.

[157] *Furnarius*, the ovenbird, the genus which gives its name to the family Furnariidae. Found from southern Mexico to Patagonia the family shows the greatest measure of diversity in the southern part of its range. Darwin collected a number of species belonging to the family, paying particular attention to *Furnarius rufus*, the Rufous Hornero (specimen 1200) and *Geositta cunicularia*, the Common Miner (specimen 1222*). The asterisked specimen is part of the collections of the British Museum (Natural History) at Tring. The Museum also owns another Darwin specimen of the same species, unnumbered, and according to its label presented by Sir William Burnett (1779–1861, physician to William IV) and Robert Fitzroy (note 27). In the *Journal of Researches* Darwin referred to *Furnarius rufus* according to its common name as the 'Casara' or house-builder and to *Geositta cunicularia* as the 'Casarita' or little house-builder. As the similarity of the two local names suggests, the second bird is something like a smaller version of the first, though its plumage is more earth-brown and not so rufous. For further information on the birds see 'Darwin's Ornithological Notes'; *JR*, pp. 112–113, 353, and 477; and Gould, *The Zoology of the Voyage of H.M.S. Beagle. Part III: Birds*, pp. 64–65.

[158] Caracaras are large carrion-feeding birds belonging to the family Falconidae. They are very common in parts of South America, and Darwin collected a number of specimens. In his 'Ornithological Notes' Darwin mentions four species: *Polyborus plancus*, the Crested Caracara (p. 234), *Milvago chimango*, the Chimango Caracara (p. 234 top and p. 236; also see specimen 1204*), *Phalcoboenus australis*, the Striated Caracara (p. 236, specimen 1882), and *Phalcoboenus albogularis*, the White-throated Caracara (p. 238, specimen 2029). In his 'Ornithological Notes' Darwin also referred to the Galapagos hawk as a caracara (p. 238), though John Gould (note 149) later corrected him. The asterisked specimen of *M. chimango* is part of the collections at the British Museum (Natural History) at Tring. The Museum also has an unnumbered *Beagle* specimen of *M. chimango* which lacks its original label, as well as a specimen of *P. plancus* presented by William Burnett (note 157) and Robert Fitzroy (note 27) without data. For more on these birds see 'Darwin's Ornithological Notes', pp. 233–239; *JR*, pp. 63–69, 256, 461; John Gould, 'Observations on the Raptorial Birds in Mr. Darwin's Collection, with characters of the New Species', *Proceedings of the Zoological Society of London*, vol. 5 (1837), pp. 9–11; and Gould, *The Zoology of the Voyage of H.M.S. Beagle. Part III: Birds*, pp. 9–31.

[159] Calandria, *Mimus saturninus modulator*, the Chalk-browed Mockingbird which Darwin collected at Maldonado (specimen 1213). In this entry Darwin probably also

had in mind other mockingbirds he collected in South America and the Galápagos Islands. Continental forms included *Mimus patagonicus*, the Patagonian Mockingbird (specimens 1461 and 1620), and *Mimus thenca*, the Chilean Mockingbird (specimen 2169). The Galápagos forms included: *Nesomimus trifasciatus* (Gould, 1837) which Darwin collected on Charles Island (specimen 3306), *Nesomimus parvulus* (Gould, 1837) which Darwin collected at Albemarle Island (specimen 3349), and *Nesomimus melanotis* (Gould 1837) which Darwin collected at Chatham Island (specimen 3307). The British Museum (Natural History) at Tring owns these specimens which, however, no longer bear Darwin's original labels. Classification of the Galápagos forms is from Michael Harris, *A Field Guide to the Birds of Galapagos* (London, 1974), the most recent treatment of these birds. The only point bearing on Darwin's specimens where Harris's grouping of the birds differs from that of Gould is with respect to the mockingbird of James Island. Harris places the mockingbirds on James Island with those of Albemarle Island; Gould placed them with the group on Chatham Island. For further discussion of all six mockingbirds described in this note, including plates on the three Galápagos species, see Gould, *The Zoology of the Voyage of H.M.S. Beagle. Part III: Birds*, pp. 60–64. Also see 'Darwin's Ornithological Notes'; *JR*, pp. 62–63, 461; and Gould's report on the three Galápagos species in *Proceedings of the Zoological Society of London*, vol. 5 (1837), p. 27.

160 The 'C' of 'Crust' is written over an 'f'.

161 William J. Burchell (1782–1863), English naturalist; personal communication. Burchell's *Travels in the Interior of Southern Africa*, vol. 2, p. 207 is quoted on the subject of the large size of South African animals compared to animals from other continents in *JR*, p. 101.

162 William J. Burchell (note 161), personal communication. See *GSA*, p. 3: "Mr. Burchell informs me, that he collected at Santos (lat. 24° S.) oyster-shells, apparently recent, some miles from the shore, and quite above the tidal action." During his South American travels of 1825–1829 Burchell made extensive zoological and botanical collections but never published significantly on them in later life. In this paragraph Burchell's name is written in light brown ink above the line, which would indicate a later dating than other entries on the page.

163 Capt. Robert Fitzroy (note 27), personal communication. See also *JR*, pp. 266–267: "I have heard Captain FitzRoy remark, that on entering any of these channels [at Tierra del Fuego] from the outer coast, it is always necessary to look out directly for anchorage; for further inland the depth soon becomes extremely great."

164 Charles Lyell (note 3), personal communication. The reference is to Leopold von Buch, *Description physique des îles Canaries* (Paris, 1836), p. 428: "Ces émanations sulfureuses paraissent donner aux volcans de Java un caractère tout particulier qui n'appartient certainement pas avec le même degré d'intensité et de fréquence à la plupart des autres volcans de la surface du globe." See *GSA*, pp. 238–239.

[165] See Edward Kendal, 'Account of the Island of Deception, one of the New Shetland Isles. Extracted from the private Journal of Lieutenant Kendal, R.N., embarked on board his Majesty's sloop *Chanticleer*, Captain Forster, on a scientific voyage...', *Journal of the Royal Geographical Society*, vol. 1 (1832), p. 64: "Possession Cape is situated in 63° 46′ S., and 61° 45′ W. We procured specimens of its rock...." Also p. 63 where the land is described as being composed "of a collection of needle-like pinnacles of sienite." Capt. Henry Foster, F.R.S. (1796–1831) commanded the *Chanticleer* from 1828–1831, Darwin's misspelling of his name deriving from an identical misspelling in the title of the article cited here.

[166] Darwin's estimate of the dimensions of Deception Island is taken from the map facing p. 64 of Kendal, 'Account of the Island of Deception' (note 165).

[167] See Edward Kendal, 'Account of the Island of Deception' (note 165), p. 65: "There was nothing in the shape of vegetation except a small kind of lichen, whose efforts are almost ineffectual to maintain its existence amongst the scanty soil afforded by the penguins' dung." P. 66: "Having observed a mound on the hill immediately above this cove, and thinking that something of interest might be deposited there, I opened it; and found a rude coffin, the rotten state of which bespoke its having been long consigned to the earth, but the body had undergone scarcely any decomposition. The legs were doubled up, and it was dressed in the jacket and cap of a sailor, but neither they nor the countenance were similar to those of an Englishman." Also p. 66: "We took the hint of the freezing over of the cove, and effected our retreat.... We quitted it on the 8th of March...." See *JR*, p. 613.

[168] James Cook, *A Voyage towards the South Pole, and round the World...1772–1775* (London, 1777), vol. 2. There is, facing p. 177, a full page map of Christmas Sound with numerous soundings included. On p. 200 Cook commented of the entire south-western coast of Tierra del Fuego: "For to judge of the whole by the parts we have sounded, it is more than probable that there are soundings all along the coast, and for several leagues out to sea. Upon the whole, this is, by no means, the dangerous coast it has been represented."

[169] James Cook, *A Voyage to the Pacific Ocean. Undertaken...for Making Discoveries in the Northern Hemisphere...1776–1780* (London, 1784), vol. 1, pp. 78–79 records that at Kerguelen Land: "A prodigious quantity of seaweed grows all over it, which seemed to be the same sort of weed that Mr. Banks distinguished by the name of *fucus giganteus*. Some of this weed is of a most enormous length, though the stem is not much thicker than a man's thumb. I have mentioned, that on some of the shoals upon which it grows, we did not strike ground with a line of twenty-four fathoms. The depth of water, therefore, must have been greater. And as this weed does not grow in a perpendicular direction, but makes a very acute angle with the bottom, and much of it afterwards spreads many fathoms on the surface of the sea, I am well warranted to say, that some of it grows to the length of sixty fathoms and

upward." See *JR*, pp. 303–304, where Darwin quoted from this passage but erroneously credited it to the narrative of Cook's second rather than his third voyage. In Darwin's notebook entry the expression '$\underset{\cdot}{\underline{24}}$' would seem to be a variant of '$\dot{\overline{24}}$'. See note 25.

[170] Benjamin Bynoe (1804–1868), Assistant and later Acting Surgeon aboard H.M.S. *Beagle*; personal communication. From the use of the present tense in this entry it would seem that Darwin saw or corresponded with Bynoe after the voyage. If so, this would not be the first occasion on which Darwin discussed geological issues with Bynoe. See, for example, Darwin MSS, Cambridge University Library, vol. 34 (ii), fol. 182 for Darwin's notes on a conversation with Bynoe during the voyage on geological topics.

[171] Woodbine Parish (note 143), personal communication. Later published in Parish, *Buenos Ayres and the Provinces of the Rio de la Plata* (London, 1839), p. 242: "It is related that for many years after its foundation, the inhabitants [of Córdoba] were subjected to much inconvenience from the occasional overflowings of a lake in the neighbouring hills, until an earthquake swallowed up its waters, and drained it apparently forever."

[172] Sir Roderick Impey Murchison, F.R.S. (1792–1871), British geologist, fellow of the Geological Society of London, twice its president (1831–1833; 1841–1843) and in 1837 a vice-president; personal communication. In the 1830s Murchison was engaged in his great work on the stratigraphy of palaeozoic rocks, which culminated in his identification of the Silurian system, which he named and described. See Murchison, *The Silurian System* (London, 1839), chapter 18, pp. 216–222 on "Lower Silurian Rocks.—3rd Formation of 'Caradoc Sandstone'." Also p. 583, "There is... a phenomenon of the highest importance, connected with the distribution of organic remains in the older strata, which has not been adverted to; namely, that the same forms of crustaceans, mollusks and corals, are said to be found in rocks of the same age, not only in England, Norway, Russia, and various parts of Europe, but also in Southern Africa, and even at the Falkland Islands, the very antipodes of Britain. This fact accords, indeed, with what has been ascertained concerning the wide range of animal remains in deposits equivalent to our oolite and lias; for in the Himalaya Mountains, at Fernando Po, in the region north of the Cape of Good Hope, and in the Run of Cutch and other parts of Hindostan, fossils have been discovered, which, as far as the English naturalists who have seen them can determine, are undistinguishable from certain oolite and lias fossils of Europe." To this remark Murchison added in a footnote: "The fossils from the Falkland Islands were discovered by Mr. C. Darwin, and they appear to me to belong to the Lower Silurian Rocks." Also see *JR*, p. 253.

[173] Rev. William Daniel Conybeare, F.R.S. (1787–1857), English geologist, early member (1811) of the Geological Society of London; later dean of Llandaff. In his

'Report on the Progress, Actual State, and Ulterior Prospects of Geological Science' (*Report of the First and Second Meetings of the British Association for the Advancement of Science* [London, 1833], p. 396), Conybeare had expressed a high opinion of *Silliman's Journal* as a source for North American geology. This journal, formally entitled the *American Journal of Science and the Arts*, contained the following full-length articles on North American geology for the year 1835: (vol. 27) Julius T. Ducatel and John H. Alexander, 'Report on a projected Geological and Topographical Survey of the State of Maryland', pp. 1–38; A. B. Chapin, 'Junction of Trap and Sandstone, Wallingford, Conn.', pp. 104–112; Henry D. Rogers, 'On the Falls of Niagara and the reasonings of some authors respecting them', pp. 326–335; 'Notice of the Transactions of the Geological Society of Pennsylvania, Part I', pp. 347–355; Charles U. Shepard, 'On the Strontianite of Schoharie, (N.Y.) with a Notice of the Limestone Cavern in the same place', pp. 363–370; (vol. 28) John Ball, 'Geology, and physical features of the country west of the Rocky Mountains, &c.', pp. 1–16; T. A. Conrad, 'Observations on the Tertiary Strata of the Atlantic Coast', pp. 104–111, 280–282; John Gebhard, 'On the Geology and Mineralogy of Schoharie, N. Y.', pp. 172–177; Samuel George Morton, 'Notice of the fossil teeth of Fishes of the United States, the discovery of the Galt in Alabama, and a proposed division of the American Cretaceous Group', pp. 276–278; and Joseph G. Totten, 'Descriptions of some Shells, belonging to the Coast of New England', pp. 347–353. Briefer reports on aspects of North American geology appear under the heading of 'Miscellanies—Foreign and Domestic' in both volumes.

[174] Review of '*A Collection of Memoirs and Documents Relative to the History, Ancient and Modern, of the Provinces of the Rio de la Plata.*—[*Coleccion de obras, &c.*] by Pedro de Angelis', *Athenæum*, no. 496 (29 April 1837), p. 302: "La Cruz [Luis de la Cruz] volunteered to conduct the expedition [for the purpose of surveying a carriage road between Concepción and Buenos Aires] at his own expense, and being accompanied by some Chilian traders, well acquainted with the Pampas, and also by some caciques of the Pehuenche Indians, he started from the fort of Ballenar, near the volcano of Antuco, in the Andes, in the beginning of April—the autumn of that climate...The length of the road which he surveyed, and actually measured with the chain, was 172 Spanish leagues and a few yards [894.4 km (555 miles)]. The expense of rendering it practicable for carts was estimated by him at 46,000 pesos, the greater part of which sum was required for the passage through the mountains. In many places the large stones which covered the ground were to be cleared away; but the chief obstacles were the cracked streams of lava to be crossed in the Andes, and the numerous banks of rough scoriæ or ashes occurring in the plains as well as the mountains." Darwin misdated his reference to this review in the *Athenæum* by a year.

[175] Pedro de Angelis, *Coleccion de obras y documentos relativos a la historia antigua y moderna de las provincias del Rio de la Plata* (Buenos Aires, 1836–1837), 6

vols. Darwin's reference is to the first two volumes of this series which were published in 1836.

[176] Angelis, *Coleccion de obras* (note 175). Woodbine Parish (note 143) was certainly mentioned in this context because of his association with Buenos Ayres and the United Provinces of La Plata. Parish would have been a likely owner, and thus a possible lender, of Angelis's work.

[177] [W. D. Cooley], Review of *Coleccion de obras y documentos relativos a la historia antigua y moderna de las provincias del Rio de la Plata, ilustrados con notas y disertaciones* by Pedro de Angelis, *Edinburgh Review*, vol. 65 (1837), pp. 87–109. *The Wellesley Index to Victorian Periodicals, 1824–1900* is the source of the reviewer's name. The 'March 1835' notation in this entry is puzzling since the date is rather far removed from either the date of publication of Angelis's work or the date of the 'present Edinburgh'.

[178] Woodbine Parish (note 143), personal communication. The distance between Quilmes and Punta Indio is approximately 70 miles (112.63 km). The two points are found along the coastline south of Buenos Aires. In his book Parish discussed a larger area covered by beds of sea shells beginning at Santa Fé two hundred and forty miles northwest of Buenos Aires (*Buenos Ayres*, p. 168): "Travelling south from Santa Fé, along the shores of the Plata, which bounds these pampas on the east, we find, at distances varying from one to six leagues inland from the river, and from fifty to one hundred and fifty miles from the sea, large beds of marine shells, which the people of those parts quarry for lime. From these deposits I have myself specimens of *Voluta Colocynthis, Voluta Angulata, Buccinum Globulosum, Buccinum Nov. Spe., Oliva Patula; Cytheræa Flexuosa*? *Mactra*? *Venus Flexuosa, Ostrea*, &c." Also see Darwin, *GSA*, pp. 2–3 for lists of shells collected along the coastline near Buenos Aires by Parish and described by Alcide Dessalines d'Orbigny (note 149) and the conchologist and fellow of the Linnean Society of London, George Brettingham Sowerby (2nd) (1812–1884).

[179] James de Carle Sowerby (1787–1871), accomplished fossil conchologist, a fellow of the Zoological Society of London and the Linnean Society of London; personal communication. See *JR*, p. 253: "Mr. Murchison, who has had the kindness to look at my specimens [of fossil shells from the Falkland Islands], says that they have a close general resemblance to those belonging to the lower division of his Silurian system; and Mr. James Sowerby is of [the] opinion that some of the species are identical." For a complete description of one group of the shells see John Morris and Daniel Sharpe, 'Description of Eight Species of Brachiopodous Shells from the Palæozoic Rocks of the Falkland Islands, *Quarterly Journal of the Geological Society of London*, vol. 2 (1846), pp. 274–278: George Brettingham Sowerby (note 178) produced the two plates. The *Journal of the Society for the Bibliography of Natural*

History, vol. 6, pt. 6, 1974, is devoted to papers on the Sowerby family and includes: J. B. MacDonald, 'The Sowerby Collection at the British Museum (Natural History)' and R. J. Cleevely, 'A provisional bibliography of natural history works by the Sowerby family'.

[180] Charles Lyell (note 3), personal communication.

[181] James Bird, 'Observations on the Manners of the Inhabitants who occupy the Southern Coast of Arabia and Shores of the Red Sea; with Remarks on the Ancient and Modern Geography of that quarter, and the Route, through the Desert, from Kosir to Keneh', *Journal of the Royal Geographical Society of London*, vol. 4 (1834), pp. 192–206. The passage from which Darwin made the inference that the land in question had been elevated is the following (p. 202): "I visited old Kosír, six miles N.W. of the modern town. The town of old Kosír is situated on the north side of an inlet of the sea, which formerly extended westward into the land about a mile, but is now crossed by a bar of sand, that prevents the ingress of the water into the former channel. The ruins of the houses are chiefly found on the north side of the channel, which is still swampy in some parts of the bottom, where, in former times, the sea formed a kind of backwater to the point of land on which the town stood....The banks which bounded the former inlet, are formed of white calcareous tuffa and sand, as is also the whole of the shore of the Red Sea at this part. The sea appears to have gradually retired from the land, and left a considerable beach between its present limits and the base of the mountains westward."

[182] See Darwin, *VI*, pp. 120–121. Citing Leopold von Buch (note 56), among others, Darwin wrote: "Lavas are chiefly composed of three varieties of feldspar, varying in specific gravity from 2.4 to 2.74; of hornblende and augite, varying from 3.0 to 3.4; of olivine, varying from 3.3 to 3.4; and lastly, of oxides of iron, with specific gravities from 4.8 to 5.2. Hence crystals of feldspar, enveloped in a mass of liquefied, but not highly vesicular lava, would tend to rise to the upper parts; and crystals or granules of the other minerals, thus enveloped, would tend to sink... Trachyte, which consists chiefly of feldspar, with some hornblende and oxide of iron, has a specific gravity of about 2.45; whilst basalt composed chiefly of augite and feldspar, often with much iron and olivine, has a gravity of about 3.0. Accordingly we find, that where both trachytic and basaltic streams have proceeded from the same orifice, the trachytic streams have generally been first erupted, owing, as we must suppose, to the molten lava of this series having accumulated in the upper parts of the volcanic focus....As the later eruptions, however, from most volcanic mountains, burst through their basal parts, owing to the increased height and weight of the internal column of molten rock, we see why, in most cases, only the lower flanks of the central, trachytic masses, are enveloped by basaltic streams. The separation of the ingredients of a mass of lava would, perhaps, sometimes take place within the body of a volcanic mountain, if lofty and of great dimensions, instead of within the

underground focus; in which case, trachytic streams might be poured forth, almost contemporaneously, or at short recurrent intervals, from its summit, and basaltic streams from its base: this seems to have taken place at Teneriffe." To this last point Darwin added a footnote: "Consult von Buch's well-known and admirable *Description Physique* of this island [Teneriffe], which might serve as a model of descriptive geology." See von Buch, *Description physique des îles Canaries*, the entire section pp. 153–228.

[183] The 'Avestruz' was the local name for the rhea. See notes 149 and 153.

[184] 'Mr Brown' was obviously a guest with Darwin at the house of Woodbine Parish (note 143), and someone with first-hand knowledge of South American geography. Lacking a first name for Brown, and a good cross-reference, one can only speculate on his identity. He may have been William Brown (1777–1857), an admiral in the navy of Buenos Aires, a native of Ireland, and the only Brown mentioned in Nina L. Kay Shuttleworth, *A Life of Sir Woodbine Parish* (London, 1910). See also Michael G. Mulhall, *The English in South America* (Buenos Aires and London, 1878), p. 166 for information which places Brown in Ireland in 1836 and therefore plausibly in London in 1837.

[185] Woodbine Parish (note 143) referring to Anthony Zachariah Helms, *Travels from Buenos Ayres, by Potosi, to Lima* (2nd ed.; London, 1807); personal communication. Helms associated the granitic boulders he found around Potosí, Bolivia, which is situated in the Cordillera proper, with granite found in Tucumán, a province in northwestern Argentina. Parish considered Tucumán to lie in the upper parts of the Sierra de Córdoba, a low range of pampean mountains. (*Buenos Ayres*, p. 254) On the subject of the travelled boulders, see Helms, p. 45: "It in a particular manner excited my astonishment here, to find the highest snow-capt mountains within nine miles from Potosi, covered with a pretty thick stratum of granitic stones, rounded by the action of water. How could these masses of granite be deposited here, as there is a continual descent to Tucuman, where the granitic ridge ends, and from Tucuman to Potosi it consists of simple argillaceous shistus? Have they been rolled hither by a general deluge, or some later partial revolution of nature?" Darwin quoted from this passage in the *JR* (p. 290), and added, "He [Helms] supposes they [the boulders] must have come from Tucuman, which is several hundred miles distant: yet at p. 55 he says, at Iocalla (a few leagues only from Potosi), 'a mass of granite many miles in length, rises in huge weatherbeaten rocks:' the whole account is to me quite unintelligible." Unlike Darwin, Parish did not quarrel with Helms' account. See *Buenos Ayres*, p. 254.

[186] 'Signor Rozales' would also seem to have been a guest with Darwin at the house of Sir Woodbine Parish (note 143). Again, lacking a first name or a good cross-reference, one can only speculate as to his identity. From his last name and the nature of his remarks, one may presume that he was South American, likely Chilean. If he were sufficiently well known to be included in standard biographical dictionaries, he

was most likely related to Juan Enrique Rosales (d. 1825), a hero of Chile's struggle for independence. One member of that family who can definitely be placed in Europe in 1837 was Francisco Javier Rosales (d. 1875), Chilean chargé d'affaires to Paris from 1836–1853. Another member of the family probably in Europe at the time was Vicente Pérez Rosales (1807–1886), subsequently a well-known author and colonization agent for the Chilean government in Europe.

[187] Edward Turner, F.R.S. (1798–1837), the chemist, as quoted in Thomas Allan, 'On a Mass of Native Iron from the Desert of Atamaca [sic] in Peru', *Transactions of the Royal Society of Edinburgh*, vol. 11 (1831), p. 226: "Externally it [the specimen] has all the characters of meteoric iron. The metal in the specimen is tough, of a whiter colour than common iron, and is covered on most parts with a thin film of the oxide of iron. The interstices contain olivine." The proportions of iron, nickel and cobalt in the specimen are given as follows (p. 228):

Iron	93.4
Nickel	6.618
Cobalt	0.535
	100.553

Undoubtedly Darwin obtained this reference by way of Woodbine Parish. See Parish, *Buenos Ayres*, pp. 257–263 for a discussion of the specimen, which Parish had collected, and of Turner's conclusions. Parish doubted the meteoric origin of the specimen.

[188] For such a map see Alexander von Humboldt, *Atlas géographique et physique des régions equinoxiales du nouveau continent* (Paris [F. Schoell], 1814), plate 5 entitled "Esquisse hyposométrique des nœuds de montagnes et des ramifications de la Cordillère des Andes depuis le Cap de Horn jusqu'a l'Isthme de Panama..." The library of the Geological Society of London does not presently hold a copy of this atlas, although, according to the librarian, it once may have. It does hold a presentation copy of the first four volumes of an octavo edition of Humboldt's voyage published in Paris by Librairie grecque-latine-allemande. Volumes 1 and 2 are dated 1816; volumes 3 and 4, 1817. The title pages of these volumes refer to an accompanying atlas, but, from the evidence of library catalogues, it is questionable whether one was published specifically for this edition.

[189] Woodbine Parish (note 143), personal communication. Parish did not include this account in *Buenos Ayres*.

[190] Edmond Temple, *Travels in Various Parts of Peru, Including A Year's Residence in Potosi*, vol. 1, p. 116: "[January] 19th [1826], when about to rise with the sun, as was our custom, we suddenly felt ourselves shaken in our beds, and thought it

was occasioned by a dog or a pig, frequent visitors prowling about for the fragments of the last meal; we therefore all, at the same moment, looked under our beds, with the intention of chasing away the intruder." And, p. 146: "Did you feel the earthquake?—At what hour?—Where were you at the time?—What did you fancy?—What did you do?—These are questions I am putting to every body I chance to converse with, and I do not think I ever felt greater interest on any subject than in the various accounts I hear respecting this phenomenon." Temple's description of his route (p. 109) places him in the province of Santiago del Estero, just over the border of the province of Córdoba, at the time when the earthquake occurred. For the account of an earthquake in Córdoba causing the disappearance of a lake see note 171.

191 Arsène Isabelle, *Voyage à Buénos-Ayres et à Porta-Alègre...1830–1834* (Havre, 1835), pp. 454–455: "Au nord-est du *passo*, à distance de quatre à cinq lieues, est une montagne boisée, appelée *Serra do Butucarahy*, s'étendant un peu à sa base, à l'est et à l'ouest, formant comme un chaînon de monts élevés indépendans de la *Serra-Grande*, et d'ailleurs placé dans une direction parallèle à celle-ci. Vue de loin (on l'aperçoit du Jacuy), elle ne paraît être qu'un pic très élevé, mais en approchant on voit que le mamelon du centre se termine par une plate forme assez grande. Je suis porté à croire que cette montagne est volcanique, parce queles *moradores* du lieu m'ont assuré avoir entendu des détonations très fortes dans son intérieur; ils prétendent encore qu'il y a un lac à la cime, dont les eaux, en filtrant ou en débordant, produisent des éboulemens qui mettent à nu la roche qu'elle semble avoir pour noyau; aussi la partie supérieure est—elle devenue inaccessible à cause de sa dénudation. Après les grandes pluies d'orage, et pendant les gelées, l'eau se trouvant dans les fissures du rocher en détache des fragmens qui tombent avec fracas; sa grande hauteur, ou plutôt son isolement attire le tonnerre, ce qui fait que cette montagne est souvent foudroyée." The hill described is probably Coxilha which lies to the northeast of the Rio Botucaraí [30° 0′ S., 52° 46′ W.] in Brazil. It is not an active volcano, nor are there any in the area.

192 Jean Baptiste Joseph Boussingault, 'Sur les tremblemens de terre des Andes', *Bulletin de la Société géologique de France*, vol. 6 (1834–1835), pp. 54–56, as cited in Charles Lyell, *Principles of Geology* (4th ed.; London, 1835), vol. 2, p. 96: "In Quito, many important revolutions in the physical features of the country are said to have resulted, within the memory of man, from the earthquakes by which it has been convulsed. M. Boussingault declares his belief, that if a full register had been kept of all the convulsions experienced here and in other populous districts of the Andes, it would be found that the trembling of the earth had been incessant. The frequency of the movement, he thinks, is not due to volcanic explosions, but to the continual falling in of masses of rock which have been fractured and upheaved in a solid form at a comparatively recent epoch. According to the same author, the height of several mountains of the Andes has diminished in modern times." This passage also occurred in the 5th or (March) 1837 edition of the *Principles* (vol. 2, p. 44), where Darwin may

have encountered it first in print. Darwin did not own a personal copy of the 4th edition.

[193] George Steuart Mackenzie (note 110) as quoted in John Barrow, Jr., *A Visit to Iceland...1834* (London, 1835), p. 224: "This supposition [of lava blistering, see note 110] would appear to afford a better solution of the difficult problem of accounting for those blocks of lava that are perched on high ridges, than that given by Sir George Mackenzie, who imagines this lava to have flowed from the lower ground, and calls it the 'ascending lava.' He says—'It is caused by the formation of a crust on the coating of the surface, and a case or tube being thus produced, the lava runs in the same manner as water in a pipe.'" The quotation is from Mackenzie, *Travels in the Island of Iceland*, p. 108.

[194] Sir Henry Holland, F.R.S. (1788–1873), fashionable London physician, traveller, essayist, and a distant relative to Charles Darwin through Josiah Wedgwood the potter. Holland accompanied Sir George Steuart Mackenzie (note 110) to Iceland in 1810 and was the author of the 'Preliminary Dissertation on the History and Literature of Iceland' in Mackenzie, *Travels in the Island of Iceland*, pp. 1–70. Apparently Darwin intended to consult him on the subject of blistered lava. On this see Barrow, *A Visit to Iceland*, p. 223: "Dr. Holland, in his account of the Mineralogy of Iceland, seems to countenance the opinion of these masses having been thrown up on the very spot they occupy, observing there was one formation of lava which had every appearance of not having flowed. Speaking of these masses of lava, he says: — 'It was heaved up into large bubbles or blisters, some of which were round, and from a few feet to forty or fifty in diameter; others were long, some straight, and some waved. A great many of these bubbles had burst open, and displayed caverns of considerable depth.'" However, this description, which Barrow attributed to Holland, is rather to be found in Mackenzie's chapter entitled 'Mineralogy' in *Travels in the Island of Iceland*, p. 390. Barrow's error seems to have stemmed from a mistaken belief that Holland rather than Mackenzie wrote the chapter on mineralogy. See also *VI*, pp. 95–96, and 103.

[195] Barrow, *A Visit to Iceland*, pp. 276–277: "Here, then, we have the plain and undeniable evidence of subterranean or sub-marine fire, exerting its influence under the sea, almost in a direct line, to the extent of 16 1/2 degrees of latitude, or more than 1100 statute miles. If we are to suppose that one and the same efficient cause has been exerted in heaving up this extended line of igneous formations, from Fairhead to Jan Meyen, we may form some vague notion how deep-seated the fiery focus must be to impart its force, perhaps through numerous apertures, in a line of so great an extent, and nearly in the same direction. It may probably be considered the more remarkable, that no indication whatever is found of volcanic fire on the coast-line of Old Greenland, close to the westward of the last-mentioned island, and also to Iceland, nor on that of Norway on the opposite side, nor on that of Spitzbergen; on these places all is granite, porphyry, gneiss, mica-slate, clay-slate, lime, marble, and sandstone."

[196] 'Bosh' is written in the margin in pencil. Other entries on the page are in ink.

[197] Alexander von Humboldt, *Political Essay on the Kingdom of New Spain* (London, 1811), vol. 3, p. 113. Quoted correctly with minor variations in capitalization, punctuation, and the insertion of '&' for 'and'.

[198] Robert Brown, F.R.S. (1773–1858), pre-eminent British botanist of his day, from 1806 to 1822 librarian to and thereafter a fellow of the Linnean Society of London. In 1827 Brown arranged for the transfer of the botanical collection of Sir Joseph Banks, F.R.S. (1743–1820) to the British Museum, and from 1827 to his death Brown supervised the botanical collections of the Museum. Brown also assembled a valuable collection of fossil woods ('F.W.') which he bequeathed to the Museum.

[199] Darwin was referring here to the opinion of James Hutton (note 130) respecting the formation of fossil wood. In Hutton's view 'undulations' in silicified fossil wood would be traced to the action of exterior heat and pressure. See Playfair, *Illustrations of the Huttonian Theory of the Earth*, pp. 24–25: "...wherever they [fossils] bear marks of having been fluid, these marks are such as characterize the fluidity of fusion [caused by igneous consolidation], and distinguish it from that which is produced by solution in a menstruum.... Fossil-wood, penetrated by siliceous matter, is a substance well known to mineralogists; it is found in great abundance in various situations, and frequently in the heart of great bodies of rock. On examination, the siliceous matter is often observed to have penetrated the wood very unequally, so that the vegetable structure remains in some places entire; and in other places is lost in a homogeneous mass of agate or jasper. Where this happens, it may be remarked, that the line which separates these two parts is quite sharp and distinct, altogether different from what must have taken place, had the flinty matter been introduced into the body of the wood, by any fluid in which it was dissolved, as it would then have pervaded the whole, if not uniformly, yet with a regular gradation. In those specimens of fossil-wood that are partly penetrated by agate, and partly not penetrated at all, the same sharpness of termination may be remarked, and is an appearance highly characteristic of the fluidity produced by fusion."

[200] Humboldt, *Political Essay on the Kingdom of New Spain*, vol. 3, p. 113. The original quotation begins, 'The true native iron,...' and varies slightly from Darwin's citation in punctuation and in the spelling out of the word 'and'.

[201] Humboldt, *Political Essay on the Kingdom of New Spain*, vol. 3, pp. 129–130: "The Mexican veins are to be found for the most part in *primitive* and *transition* rocks ...and rarely in the rocks of *secondary* formation...In the old continent *granite*, *gneiss* and *micaceous slate* (*glimmer-schiefer*) constitute the crest of high chains of mountains. But these rocks seldom appear outwardly on the ridge of the Cordilleras of America, particularly in the central part contained between the 18° and 22° of north latitude. Beds of amphibolic porphyry, greenstone, amygdaloid, basalt and

other trap formations of an enormous thickness cover the granite and conceal it from the geologist. The coast of Acapulco is formed of granite rock. Ascending towards the table land of Mexico we see the granite pierce through the porphyry for the last time between Zumpango and Sopilote. Farther to the east in the province of Oaxaca the granite and gneiss are visible in table lands of considerable extent traversed by veins of gold."

[202] Humboldt, *Political Essay on the Kingdom of New Spain*, vol. 3, p. 131. The original sentence reads "The *porphyries...*"

[203] Humboldt, *Political Essay on the Kingdom of New Spain*, vol. 3, pp. 131–132: "They [the Mexican porphyries] are all characterized by the constant presence of amphibole and the absence of quartz, so common in the primitive porphyries of Europe, and especially in those which form beds in gneiss. The *common felspar* is rarely to be seen in the Mexican porphyries; and it belongs only to the most antient formations, those of Pachuca, Real del Monte and Moran, where the veins furnish twice as much silver as all Saxony. We frequently discover only *vitreous felspar* in the porphyries of Spanish America. The rock which is intersected by the rich gold vein of Villalpando near Guanaxuato is a porphyry of which the basis is somewhat a kin to *klingstein* (*phonolite*), and in which amphibole is extremely rare. Several of these parts of New Spain bear a great analogy to the problematical rocks of Hungary, designated by M. Born by the very vague denomination of *saxum metalliferum*. The veins of Zimapan which are the most instructive in respect to the theory of the stratification of minerals are intersected by porphyries of a *greenstone* base which appear to belong to trap rocks of new formation. These veins of Zimapan offer to oryctognostic collections a great variety of interesting minerals such as the fibrous zeolith, the stilbite, the grammalite, the pyenite, native sulphur, spar fluor, baryte suberiform asbestos, green grenats, carbonate and chromate of lead, orpiment, chrysoprase, and a new species of opal of the rarest beauty, which I made known in Europe, and which M. M. Karsten and Klaproth have described under the name of (*feuer-opal*)."

[204] Humboldt, *Political Essay on the Kingdom of New Spain*, vol. 3, pp. 133–134: "In proportion as the north of Mexico shall be examined by intelligent geologists, it will be perceived that the metallick wealth of Mexico does not exclusively belong to primitive earths and mountains of transition, but extend also to those of *secondary formation*. I know not whether the lead which is procured in the eastern parts of the intendancy of San Luis Potosi is found in veins or beds, but it appears certain, that the veins of silver of the real de Catorce, as well as those of the Doctor and Xaschi near Zimapan, traverse the *alpine lime-stone* (*alpenkalkstein*); and this rock reposes on a *poudingue* with silicious cement which may be considered as the most antient of secondary formations. The alpine lime-stone and the jura lime-stone (*jurakalkstein*) contain the celebrated silver mines of Tasco and Teuilotepec in the intendancy of Mexico; and it is in these calcareous rocks that the numerous veins which in this

country have been very early wrought, display the greatest wealth. . . . The result of this general view of the metalliferous depositories (erzführende lagerstätte) is that the cordilleras of Mexico contain veins in a great variety of rocks, and that those rocks which at present furnish almost the whole silver annually exported from Vera Cruz, are the *primitive slate*, the *grauwakke*, and the *alpine lime-stone*, intersected by the *principal veins* of Guanaxuato, Zacatecas and Catorce."

[205] Friedrich Hoffmann, *Geschichte der Geognosie* (Berlin, 1838), the section 'Dämpfe verändern die vulkanischen Gesteine', pp. 480–481. This entry is written in small handwriting in light brown ink, as are all other bracketed entries on page 165e.

[206] Humboldt, *Political Essay on the Kingdom of New Spain*, vol. 3, p. 128: "How can he [the naturalist] draw general results from the observation of a multitude of small phenomena [regarding metalliferous deposits], modified by causes of a purely local nature, and appearing to be the effects of an action of chemical affinities, circumscribed to a very narrow space?"

[207] Eilhert Mitscherlich, 'On Artificial Crystals of Oxide of Iron', *Edinburgh Journal of Natural and Geographic Science*, vol. 2 (1830), p. 302: "So greatly do these [crystals of oxide of iron in a pottery furnace] resemble the crystals [of specular iron] from volcanoes, that the same theory of formation may be applied to both. The first are formed in a pottery furnace, in which the vessels, when baked, are glazed by means of common salt. The clay used consists principally of silica, alumina, and a little oxide of iron. The salt is volatilized, and water coming in contact with the surface of the vessels, new compounds are produced, the water is decomposed, muriatic acid is formed, and the soda produced unites with the silica to make the necessary glass. As to the oxide of iron, its history will be best understood by an experiment or two. If a mixture of salt, oxide of iron, and silica, be heated to redness in a tube, and water in vapour be passed over it, much muriatic acid is formed, but very little chloride of iron, and crystallized oxide of iron will be found in the mass: but if muriatic acid be brought in contact with ignited oxide of iron, water and chloride of iron are formed, and sublime; if the chloride of iron come in contact with more water, muriatic acid is first developed, then chloride of iron, and a residue of crystallized oxide of iron remains. The formation of chloride of iron by the action of muriatic acid upon oxide of iron appears, therefore, to depend upon the proportion of water present. M. Mitscherlich applies these experiments and principles in explanation of the manner in which volcanic crystallized oxide of iron is formed —all the conditions necessary, according to the above view, being present in those cases, where heretofore it had been supposed the oxide of iron, as such, had been actually sublimed."

[208] Humboldt, *Political Essay on the Kingdom of New Spain*, vol. 3, p. 134, "Thus it is in a *primitive slate* (*ur-thon schiefer*) on which a clayey porphyry containing grenats reposes, that the wealth of *Potosi* in the kingdom of Buenos-Ayres is contained.

On the other hand, in Peru the mines of Gualgayoc or Chota and that of Yauricocha or Pasco which together yield annually double the quantity of all the German mines, are found in an *alpine lime-stone*. The more we study the geological constitution of the globe on a large scale the more we perceive that there is scarcely a rock which has not in certain countries been found eminently metalliferous. The wealth of the veins is for the most part totally independent of the nature of the beds which they intersect." And pp. 142–143: "The province of Quito, and the Eastern part of the kingdom of New Granada, from the 3° of South latitude, to the 7° of North latitude; the Isthmus of Panama, and the mountains of Guatimala, contain for a length of 600 leagues, vast extents of ground in which no vein has hitherto been wrought with any degree of success. It would not, however, be accurate to advance that these countries which have in a degree, been convulsed with volcanos are entirely destitute of gold and silver ore. Numerous metalliferous depositories may be concealed by the super-position of strata of basalt, amygdaloid, porphyry with *greenstone* base, and other rocks comprehended by geologists, under the general name of *trapp-formation*."

[209] Humboldt, *Political Essay on the Kingdom of New Spain*, vol. 3, pp. 152–153: "In Peru, the greatest part of the silver extracted from the bowels of the earth is furnished by the *pacos*, a sort of ores of an earthy appearance, which M. [Martin Heinrich] Klaproth was so good as to analyse at my request, and which consist of a mixture of almost imperceptible parcels of native silver, with the brown oxyde of iron. In Mexico on the other hand, the greatest quantity of silver annually brought into circulation, is derived from those *ores* which the Saxon miner calls by the name of *dürre erze* especially from *sulfuretted silver*, (or vitrous *glaserz*) from *arsenical grey-copper* (*fahlerz*) and *antimony*, (*grau* or *schwarzgiltigerz*) from *muriated silver*, (*hornerz*) from *prismatic black silver*, (*spödglaserz*), and from red silver (*rothgiltigez*). We do not name native silver among these ores, because it is not found in sufficient abundance to admit of any very considerable part of the total produce of the mines of New Spain being attributed to it." Also p. 154: "The muriated silver which is so seldom found in the veins of Europe, is very abundant in the mines of Catorce, Fresnillo, and the Cerro San Pedro, near the town of San Luis Potosi. . . . In the veins of Catorce, the muriated silver is accompanied with molybated lead, (*gelb-blei-erz*) and phosphated lead (*grünblei-erz*)." And p. 155: "The true mine of *white silver* (weissgiltig-erz) is very rare in Mexico. Its variety *greyish white*, very rich in lead, is to be found however in the intendancy of Sonora, in the veins of Cosala, where it is accompanied with argentiferous *galena*, red silver, brown blende, quartz and sulfated barytes. This last substance which is very uncommon among the *gangues* of Mexico, is to be also found at the Real del Doctor, near Baranca de las Tinajas, and at Sombrerete, particularly in the mine called Campechana. Spar-fluor has been only found hitherto in the veins of Lomo del Toro, near Zimapan, at Bolaños and Guadalcazar, near Catorce."

[210] Humboldt, *Political Essay on the Kingdom of New Spain*, vol. 3, p. 156. Quoted exactly except for the abbreviation of 'and' to '&' and the deletion of a comma after 'silver'.

[211] Humboldt, *Political Essay on the Kingdom of New Spain*, vol. 3, p. 157: "*Native Silver*, which is much less abundant in America, than is generally supposed, has been found in considerable masses, sometimes weighing more than 200 killogrammes [441 lbs], in the seams of Batopilas in New Biscay. These mines, which are not very briskly wrought at present, are among the most northern of New Spain. Nature exhibits the same minerals there, that are found in the vein of Kongsberg in Norway. Those of Batopilas contain filiform dendritic and silver, which intersects with that of carbonated lime."

[212] Humboldt, *Political Essay on the Kingdom of New Spain*, vol. 3, pp. 157–158: "Native silver is constantly accompanied by *glaserz* [sulfuretted silver] in the seams of Mexico, as well as in those of the mountains of Europe...From time to time small branches, or cylindrical filaments of native silver, are also discovered in the celebrated vein of Guanaxuato; but these masses have never been so considerable as those which were formerly drawn from the mine *del Encino* near Pachuca and Tasco, where native silver is sometimes contained in folia of selenite."

[213] Humboldt, *Political Essay on the Kingdom of New Spain*, vol. 3, p. 158: "A great part of the silver annually produced in Europe, is derived from the *argentiferous sulfuretted lead* (*silberhaltiger bleiglanz*) which is sometimes found in the veins which intersect *primitive and transition mountains*, and sometimes on particular *beds* (erzflöze) in rocks of *secondary formation*. In the kingdom of New Spain, the greatest part of the veins contain very little argentiferous galena; but there are very few mines in which lead ore is a particular object of their operations."

[214] Humboldt, *Political Essay on the Kingdom of New Spain*, vol. 3, p. 159: "A very considerable quantity of silver is produced from the smelting of the martial pyrites (*gemeine schwefelkiese*) of which New Spain sometimes exhibits varieties richer than the *glaserz* itself...It is a very common prejudice in Europe, that great masses of native silver are extremely common in Mexico and Peru...Also pp. 160–161: "It appears that at the formation of veins in every climate, the distribution of silver has been very unequal; sometimes concentrated in one point, and at other times disseminated in the *gangue*, and allied with other metals." And p. 162: "Although the New Continent, however, has not hitherto exhibited native silver in such considerable blocks as the Old, this metal is found more abundantly in a state of perfect purity in Peru and Mexico, than in any other quarter of the globe."

[215] Humboldt, *Political Essay on the Kingdom of New Spain*, vol. 3, p. 176: "What is the nature of the *metalliferous depository*, which has furnished these immense riches, and which may be considered as the Potosi of the northern hemisphere? What is the

position of the rock which crosses the veins of Guanaxuato? These questions are of so great importance that I must here give a geological view of so remarkable a country. The most ancient rock known in the district of Guanaxuato, is the *clay slate* (*thon schiefer*).... It is of an ash-grey or greyish-black frequently intersected by an infinity of small quartz veins, which frequently pass into talk-state [sic] (*talk schiefer*) and into *schistous chlorite*." Also, pp. 177–178: "On digging the great pit (*tiro general*) of Valenciana, they discovered banks of *syenite* of *Hornblend slate* (*Hornblend schiefer*) and true serpentine, altering with one another, and forming *subordinate beds*, in the *clay slate*.... These strata [of clay slate] are very regularly *directed* h. 8 to 9 of the miner's compass; they are inclined from 45 to 50 degrees to the south west.... Two very different formations repose on the *clay slate*: the one of porphyry...and the other, of old *freestone* in the ravins, and table lands of small elevation." And pp. 179–180: "This porphyry...is generally of a greenish colour.... The most recent [beds]... contain vitreous felspar, inchased in a mass, which sometimes passes into the petrosilex jadien, and sometimes into the pholonite [sic] or *klingstein* of Werner.... All the porphyries of the district of Guanaxuato possess this in common, that amphibole is almost as rare in them as quartz and mica."

216 Humboldt, *Political Essay on the Kingdom of New Spain*, vol. 3, pp. 180–183: "This free-stone (*urfelsconglomerat*) is a brescia with clayey cement, mixed with oxide of iron, in which are imbedded *angulous* fragments of quartz, Lydian stone, syenite, porphyry, and splintery hornstone.... This formation of old free-stone is the same with that which appears at the surface in the plains of the river Amazon, in South America.... We must not confound the brescia which contains imbedded fragments of primitive and transition rock, with another freestone, which may be designated by the name of *felspar agglomeration*.... This agglomeration...is composed of grains of quartz, small fragments of slate, and felspar chrystals, partly broken, and partly remaining untouched.... Probably the destruction of porphyries has had the greatest influence on the formation of this *felspar* freestone. It contrasts with the freestone of the Old Continent, in which some chrystals of grenats and amphibole have been found, but never...felspar in any abundance. The most experienced mineralogist, after examining the position of the *lozero* [agglomeration] of Guanaxuato, would be tempted to take it at first view, for a porphyry with clayey base, or for a porphyritic brescia (*trümmer-porphyr*).... These formations of old *freestone* of Guanaxuato, serve as bases to other secondary beds, which in their *position*, that is to say in *the order of their superposition*, exhibit the greatest analogy with the secondary rocks of central Europe. In the plains of Temascatio...there is a compact limestone..."

217 Humboldt, *Political Essay on the Kingdom of New Spain*, vol. 3, p. 185: "The vein (*veta madre*) [of Guanaxuato] traverses both clay slate and porphyry. In both of these rocks, very considerable wealth has been found. Its mean direction is...[N. 52° W.] and is nearly the same with that of the *veta grande* of Zacatecas, and of the veins

of Tasco and Moran, which are all western veins (spathgänge). The inclination of the vein of Guanaxuato, is 45 or 48 degrees to the south west." Also pp. 186–187: "The *veta madre* of Guanaxuato, bears a good deal of resemblance to the celebrated vein of *Spital* of Schemnitz, in Hungary. The European miners who have had occasion to examine both these *depositories* of minerals, have been in doubt whether to consider them as true veins, or as *metalliferous beds* (*erzlager*).... If the *veta madre* was really a *bed*, we should not find *angular fragments* of its *roof* contained in its *mass*, as we generally observe on points where the *roof* is a *slate* charged with *carbone*, and the wall a talc slate. In a vein, the *roof* and the *wall* are deemed anterior to the formation of the *crevice*, and to the minerals which have successfully filled it; but a *bed* has undoubtedly pre-existed to the *strata* of the rock which compose its *roof*. [Hence] we may discover in a bed fragments of the *wall*, but never pieces detached from the *roof*."

218 Erasmus Darwin, *The Botanic Garden* (4th ed.; London, 1799), vol. 1, p. 18: "The air, like all other bad conductors of electricity, is known to be a bad conductor of heat...." Also see p. 11 on the subject of shooting stars and fireballs and pp. 249–258 for a discussion of meteors.

219 Woodbine Parish (note 143). See note 187 and see also Parish, 'Notice as to the supposed Identity of the large Mass of Meteoric Iron now in the British Museum, with the celebrated Otumpa Iron described by Rubin de Celis in the Philosophical Transactions for 1786', *Philosophical Transactions of the Royal Society of London*, vol. 124 (1834), pp. 53–54.

220 Ernst Florenz Friedrich Chladni, 'Supplément au catalogue des météores, à la suite desquels des pierres ou des masses de fer sont tombées', *Journal des mines*, vol. 26 (1809), pp. 79–80. Speaking of meteorites Chladni wrote (p. 80), "Il paraît qu'on doit aussi ranger parmi les masses dont il s'agit, celle d'un fer malléable, du poids de 97 myriagrammes, qu'un minéralogiste saxon, M. Sonnenschmidt, a trouvée dans la ville de Zacatecas, dans la Nouvelle-Espagne, où il était directeur des mines." Alexander von Humboldt (note 38) also reported the existence of this stone. See Humboldt, *Political Essay on the Kingdom of New Spain*, vol. 2, p. 293.

221 Alcide Dessalines d'Orbigny, *Voyage dans l'Amérique méridionale...1826–1833*, vol. 5, part 1 (Paris and Strasbourg, 1847), pp. 140–144 and plate 10. Individual sections of this volume were published separately earlier. According to a typewritten list compiled in 1933 by Charles Davies Sherborn of the British Museum (Natural History), the section which includes pp. 140–144 was published in 1835 and plate 10 in 1834. The three species described by d'Orbigny were *Sagitta triptera*, *Sagitta exaptera*, and *Sagitta diptera*. In this entry Darwin was noting the similarity of one of his unidentified specimens to *Sagitta triptera*. The genus *Sagitta* or 'Flèche' had been established by Jean René Constantin Quoy and Paul Gaimard in their 'Observations zoologiques faites à bord de l'Astrolabe, en Mai 1826, dans le Détroit de Gilbraltar',

Annales des sciences naturelles, vol. 10 (1827), p. 232–233. Presumably Darwin's 'additional information' on the genus appeared in his later article, 'Observations on the Structure and Propagation of the genus *Sagitta*', *Annals and Magazine of Natural History*, vol. 8 (1844), pp. 1–6 with plate.

[222] Humboldt, *Political Essay on the Kingdom of New Spain*, vol. 3, p. 189. Quoted exactly except for minor variations in punctuation, the abbreviation of 'and' to '&' and the lack of emphasis on foreign words by way of underlining.

[223] Humboldt, *Political Essay on the Kingdom of New Spain*, vol. 3, p. 205: "The *veta grande*, or principal vein [at Zacatecas], has the same direction as the *veta madre* of Guanaxuato; the others are generally in a direction from east to west." And p. 207: "This wealth is displayed...not in the ravins, and where the veins run along the gentle slope of the mountains, but most frequently on the most elevated summits, on points where the surface appears to have been tumultuously torn, in the antient revolutions of the globe."

[224] Humboldt, *Political Essay on the Kingdom of New Spain*, vol. 3, p. 210: "The greatest number of these veins [at Catorce] are *western* (*spathgänge*); and their inclination is from 25° to 30° towards the north east." P. 223: "...the vein of Moran ...inclined 84° to the north east..." P. 226: "The oldest rock which appears at the surface in this district of mines [at Tasco], is the primitive slate....Its direction is hor. 3–4; and its inclination 40° to the north-west...." Also p. 227: "The district of mines of Tasco...contains a great number of veins...all directed from the north-west to the south-east, hor. 7–9."

[225] Humboldt, *Political Essay on the Kingdom of New Spain*, vol. 3, p. 215: "What relation exists between these last beds [of porphyry], which several distinguished mineralogists consider as volcanic productions, and the porphyries of Pachuca, Real del Monte, and Moran, in which nature has deposited enormous masses of sulfuretted silver and argentiferous pyrites? This problem which is one of the most difficult in geology, will only be resolved when a great number of zealous and intelligent travellers, shall have gone over the Mexican Cordilleras, and carefully studied the immense variety of porphyries which are destitute of quartz, and which abound both in hornblend and vitreous felspar."

[226] Humboldt, *Political Essay on the Kingdom of New Spain*, vol. 3, p. 227: "These veins [in the mining districts of Tasco and the Real de Tehuilotepec], like those of Catorce, traverse both the limestone and the micaceous slate which serves for its base; and they exhibit the same metals in both rocks."

[227] Humboldt, *Political Essay on the Kingdom of New Spain*, vol. 3, p. 230: "This formation [of veins, one of four types existing at Tasco and Tehuilotepec] which is the richest of all, displays the remarkable phenomenon, that the minerals the most abundant in silver, form spheroidal balls, from ten to twelve centimetres in diameter...."

[228] Humboldt, *Political Essay on the Kingdom of New Spain*, vol. 3, pp. 347–348: "The mines of Huantajaya, surrounded with beds of rock salt are particularly celebrated on account of the great masses of native silver which they contain in a decomposed gangue; and they furnish annually between 70 and 80 thousand marcs of silver. The muriate of conchoidal silver, sulphuretted silver, galena with small grains, quartz, carbonate of lime, accompany the native silver." Also pp. 348–349: "[Antonio de] Ulloa after travelling over a great part of the Andes, affirms that silver is peculiar to the high table lands of the Cordilleras, called *Punas* or *Paramos*, and that gold on the other hand abounds in the lowest, and consequently warmest regions; but this learned traveller appears to have forgot that in Peru the richest provinces in gold are the *partidos* of Pataz and Huailas, which are on the ridge of the Cordilleras.... It [gold] has also been extracted from the right bank of the Rio de Micuipampa, between the Cerro de San Jose, and the plain called by the natives, *Choropampa* or *plain of shells*, on account of an enormous quantity of ostracites, cardium and other petrifications of sea shells contained in the formation of alpine limestone of Gualgayoc."

[229] Louis Antoine de Bougainville, *Voyage autour du monde...1766–1769*, (2nd ed.; Paris, 1772), vol. 1, p. 291: "Ces hommes bruts [the Fuegians] traitoient les chefs-d-œuvre de l'industrie humaine, comme ils traitent les loix de la nature & ses phénomènes." See *JR*, p. 242.

[230] Juan and Ulloa, *A Voyage to South America*, vol. 1, pp. 168–169: "On the coast [at Guayaquil]...is found that exquisite purple, so highly esteemed among the ancients; but the fish from which it was taken, having been either unknown or forgotten, many moderns have imagined the species to be extinct. This colour, however, is found in a species of shell-fish growing on rocks washed by the sea. They are something larger than a nut, and are replete with a juice, probably the blood, which, when expressed, is the true purple; for if a thread of cotton, or any thing of a similar kind, be dipt in this liquor, it becomes of a most vivid colour, which repeated washings are so far from obliterating, that they rather improve it; nor does it fade by wearing. ...Stuffs died with this purple are also highly valued. This precious juice is extracted by different methods. Some take the fish out of its shell, and laying it on the back of their hand, press it with a knife from the head to the tail, separating that part of the body into which the compression has forced the juice, and throw away the rest. In this manner they proceed till they have provided themselves with a sufficient quantity. Then they draw the threads through the liquor, which is the whole process. But the purple tinge does not immediately appear, the juice being at first of a milky colour; it then changes to green; and, lastly, into this celebrated purple. Others pursue a different method in extracting the colour; for they neither kill the fish, nor take it entirely out of its shell; but squeeze it so hard as to express a juice, with which they die the thread, and afterwards replace the fish on the rock whence it was taken."

[231] Juan and Ulloa, *A Voyage to South America*, vol. 1, p. 281: "As the pestilence, whose ravages among the human species in Europe, and other parts, are so dreadful,

is unknown both at Quito and throughout all America, so is also the madness in dogs. And though they have some idea of the pestilence, and call those diseases similar in their effects by that name, they are entirely ignorant of the canine madness; and express their astonishment when an European [sic] relates the melancholy effects of it."

[232] This entry is written in light brown ink.

[233] Humboldt, *Political Essay on the Kingdom of New Spain*, vol. 4, p. 58: "It is observed at Acapulco that the shakes take three different directions, sometimes coming from the west by the isthmus [which separates Acapulco from the Bay de la Langosta de la Abra de San Nicolas]...sometimes from the north west as if they were from the volcano de Colima, and sometimes coming from the south. The earthquakes which are felt in the direction of the south are attributed to submarine volcanoes; for they see here, what I often observed at night in the Callao of Lima, that the sea becomes suddenly agitated in a most alarming manner in calm and serene weather when not a breath of wind is blowing." This entry is written in light brown ink.

[234] Of the petrified trees he found on the Uspallata range Darwin wrote (*JR*, p. 406): "Mr. Robert Brown [note 198] has been kind enough to examine the wood: he says it is coniferous, and that it partakes of the character of the Araucarian tribe (to which the common South Chilian pine belongs), but with some curious points of affinity with the yew." Also see *GSA*, p. 202 for repetition of the same information. From Darwin's correspondence it is clear that Brown described the specimens of silicified wood sometime during the period from the end of March to mid-May 1837. On 28 March Darwin wrote to J. S. Henslow (note 148) telling of Brown's general interests in specimens from the *Beagle* voyage; on 10 April Darwin wrote to the English naturalist Leonard Jenyns [later Leonard Blomefield] (1800–1893): "Tell Henslow, I think my silicified wood has unflintified Mr. Brown's heart"; and on 18 May Darwin wrote to Henslow with Brown's identification of the specimens. For the Darwin–Henslow letters see Nora Barlow, ed., *Darwin and Henslow*, pp. 125, 127. For the letter to Jenyns see Francis Darwin, ed., *The Life and Letters of Charles Darwin* (3rd ed. rev.; London, 1888), vol. 1, p. 282.

[235] Jean André Deluc, *Geological Travels*. 3 vols. (London, 1810–1811).

[236] Francis Beaufort, *Karamania; or, A Brief Description of the South Coast of Asia-Minor and of the Remains of Antiquity, with Plans, Views, &c., Collected during a Survey of That Coast...in...1811–1812* (London, 1817).

[237] Ross, *A Voyage of Discovery...for the Purpose of Exploring Baffin's Bay*, Appendix No. 3, 'Table of Soundings obtained in Davis' Strait and Baffin's Bay'.

[238] William Scoresby, Jr., *An Account of the Arctic Regions, with a History and Description of the Northern Whale-fishery* (Edinburgh, 1820), vol. 1, pp. 184–194

entitled, 'Temperature, Depth, and Pressure of the Greenland Sea, with a Description of an Apparatus for bringing up Water from great Depths, and an Account of Experiments made with it'.

[239] Gilbert Farquhar Mathison, *Narrative of a Visit to Brazil, Chile, Peru, and the Sandwich Islands...1821–1822* (London, 1825).

[240] John Mawe, *Travels in the Gold and Diamond Districts of Brazil* (A new ed.; London, 1825). The first two editions of this work had been entitled *Travels in the Interior of Brazil.* This was, however, the edition which Darwin owned.

[241] Alessandro Malaspina (1754–1809), Spanish navigator of Italian birth. From 1789–1794 Malaspina led a major Spanish scientific expedition to circumnavigate the earth. However, on his return to Spain, Malaspina fell from favour, partly owing to court intrigue and partly to Malaspina's critical attitude towards Spain's treatment of her colonies. Malaspina was imprisoned and the full account of his work, which was to run to seven volumes, was never completed. His own narrative of the voyage was finally published in 1885 under the title *Viaje político-científico alrededor del mundo por las corbetas Descubierta y Atrevida...1789–1794* (Madrid, 1885). Earlier Malaspina's navigational work appeared in official Spanish charts, where it was much praised (see, for example, Parish, *Buenos Ayres*, pp. 96–98), although Malaspina was not credited for his contribution. A few of his observations were eventually published under his own name as 'Tablas de latitudes y longitudes de los principales puntos del Rio de la Plata', in Pedro de Angelis, *Coleccion de obras y documentos relativos a la historia antigua y moderna de las provincias del Rio de la Plata* (Buenos-Aires, 1837), vol. 6.

[242] Faddeĭ Faddeevich Bellinsgauzen, Двукратныя изысканія вь Южномь Ледовитомь Океанв...*1819–1821* [*Dvukratni͡a izyskanii͡a v I͡Uzhnom Ledovitom Okeane ...1819–1821*]. 2 vols. plus atlas. (St. Petersburg, 1831). The original work was published in a limited edition of 600 copies. For an English translation see Frank Debenham, ed., *The Voyage of Captain Bellingshausen to the Antarctic Seas 1819–1821.* 2 vols. (London, 1945).

[243] Otto von Kotzebue, *A Voyage of Discovery, into the South Sea and Beering's Straits...1815–1818.* 3 vols. (London, 1821).

Bibliography to the Text of the Red Notebook

Acosta, José de. *Histoire naturelle et moralle des Indes, tant Orientalles qu'Occidentalles: où, il est traicté des choses remarquables du ciel, des elemens, metaux, plantes, & animaux qui sont propres de ce pays. Ensemble de mœurs, ceremonies, loix, gouuernemens, & guerres des mesmes Indiens. Composée en castillan par Ioseph Acosta, & traduite en françois par Robert Regnault Cauxois. Dedié au Roy.* Derniere ed., reueuë & cor. de nouueau. Paris, 1600.

Alexander, James Edward. 'Notice regarding the Salt Lake Inder, in Asiatic Russia.' *Edinburgh New Philosophical Journal.* vol. 8 (1830), pp. 18–20.

Allan, Thomas. 'On a Mass of Native Iron from the Desert of Atamaca in Peru.' *Transactions of the Royal Society of Edinburgh.* vol. 11 (1831), pp. 223–228.

Allardyce, James. 'On the Granitic Formation, and direction of the Primary Mountain Chains, of Southern India.' *Madras Journal of Literature and Science.* vol. 4 (1836), pp. 327–335.

Angelis, Pedro de. *Coleccion de obras y documentos relativos a la historia antigua y moderna de las provincias del Rio de la Plata.* 6 vols. Buenos Aires, 1836–1837.

——. Review of *A Collection of Memoirs and Documents Relative to the History, Ancient and Modern, of the Provinces of the Rio de la Plata.—[Coleccion de obras, &c.]* by Pedro de Angelis. *Athenæum Journal of Literature, Science, and the Fine Arts.* no. 496 (29 April 1837), pp. 300–303.

Aubuisson de Voisins, Jean François d'. *Traité de géognosie; ou, Exposé des connaissances actuelles sur la constitution physique et minérale du globe terrestre.* 2 vols. Strasbourg, 1819.

Azara, Félix d'. *Voyages dans l'Amérique Méridionale, par don Félix de Azara, commissaire et commandant des limites espagnoles dans le Paraguay depuis 1781 jusqu'en 1801; contenant la description géographique, politique, et civile du Paraguay et de la rivière de la Plata; l'histoire de la découverte et de la conquête de ces contrées; des détails nombreux sur leur histoire naturelle, et sur les peuples sauvages qui les habitent; le récit des moyens employés par les Jésuites pour assujétir et civiliser les indigènes, etc. Publiés d'après les manuscrits de l'auteur, avec une notice sur sa vie et ses écrits, par C. A. Walckenaer; enrichis de notes par G. Cuvier, secrétaire perpétuel de la classe des sciences physiques de l'Institut, etc. Suivis de l'histoire naturelle des* *oiseaux du Paraguay et de la Plata, par le même auteur, traduite, d'après l'original espagnol, et augmentée d'un grand nombre de notes, par M. Sonnini.* 4 vols. + atlas. Paris, 1809.

Ball, John. 'Geology, and physical features of the country west of the Rocky Mountains, &c.' *American Journal of Science and Arts.* vol. 28 (1835), pp. 1–16.

Barlow, Nora, ed. *Charles Darwin's Diary of the Voyage of H.M.S. Beagle.* Cambridge, 1933.

—— ed. *Darwin and Henslow: The Growth of an Idea. Letters 1831–1860.* Berkeley and Los Angeles, 1967.

—— ed. 'Darwin's Ornithological Notes.' *Bulletin of the British Museum (Natural History)* Historical Series. vol. 2 (1963), pp. 201–278.

Barrow, Jr., John. *A Visit to Iceland, by Way of Tronyem, in the "Flower of Yarrow" Yacht, in the Summer of 1834.* London, 1835.

Beaufort, Francis. *Karamania; or, A Brief Description of the South Coast of Asia-Minor and of the Remains of Antiquity, with Plans, Views, &c. Collected during a Survey of That Coast, under Orders of the Lords Commissioners of the Admiralty, in the Years 1811 & 1812.* London, 1817.

Beechey, Frederick William. *Narrative of a Voyage to the Pacific and Beering's Strait, to Co-operate with the Polar Expeditions: Performed in His Majesty's Ship Blossom, under the Command of Captain F. W. Beechey, R.N., F.R.S. &c., in the Years 1825, 26, 27, 28.* Philadelphia, 1832.

Bellinsgauzen, Faddeĭ Faddeevich. Двукратныя изысканія въ Южномъ Ледовитомъ Океанѣ и плаваніе вокругъ свѣта, въ продолженіи 1819, 20 и 21 годовъ, совершенныя на шлюпахъ Востокѣ и Мирномъ, подъ начальствомъ Капитана Беллинсгаузэена, Командира Шлюпа Востока. Шлюпомъ Мирнымъ начальствовалъ Лейтенантъ Лаэаревъ, *etc.*

[Dvukratnyi͡a izyskanii͡a v'I͡Uahnom Ledovitom Okeane i plavanie vokrug svi͡eta, v prodolzhenii 1819, 20 i 21 godov, sovershennyi͡a na shli͡upakh Vostok i Mirnom, pod nachal'stvom Kapitana Bellinsgauzena, Komandira Shli͡upa Vostoka. Shli͡upom Mirnym nachal'stvoval Leitenant Lazarev...] 2 vols. + atlas. St Petersburg, 1831.

Bird, James. 'Observations on the Manners of the Inhabitants who occupy the Southern Coast of Arabia and Shores of the Red Sea; with Remarks on the Ancient and Modern Geography of that quarter, and the Route, through the Desert, from Kosir to Keneh.' *Journal of the Royal Geographical Society of London.* vol. 4 (1834), pp. 192–206.

Bolingbroke, Henry. *A Voyage to the Demerary, Containing a Statistical Account of the Settlements There, and of Those on the Essequebo, the Berbice, and Other Contiguous Rivers of Guyana.* London, 1807.

Bory de Saint-Vincent, Jean Baptiste. *Voyage dans les quatre principales îles des mers d'Afrique, fait par ordre du gouvernement, pendant les années neuf et dix de la République (1801 et 1802), avec l'histoire de la traversée du capitaine Baudin jusqu'au Port-Louis de l'île Maurice. Par J.B.G.M. Bory de St-Vincent, officer d'état-major; naturaliste en chef sur la corvette Le Naturaliste, dans l'expédition de découvertes commandée par le capitaine Baudin.* 3 vols. Paris, 1804.
—— ed. *Dictionnaire classique d'histoire naturelle.* 17 vols. incl. atlas. Paris, 1822–1831.

Bougainville, Louis Antoine de. *Voyage autour du monde, par le frégate du roi la Boudeuse, et la flûte l'Étoile; en 1766, 1767, 1768 & 1769.* 2 vols. 2nd ed. Paris, 1772.

Boussingault, Jean Baptiste Joseph. 'Sur les tremblemens de terre des Andes.' *Bulletin de la Société géologique de France.* vol. 6 (1834–1835), pp. 52–57.

Buch, Leopold von. *Description physique des îles Canaries, suivie d'une indication des principaux volcans du globe,...traduite de l'allemand par C. Boulanger,...revue et augmentée par l'auteur.* + atlas. Paris, 1836.

Bulkeley, John, and Cummins, John. *A Voyage to the South-Seas, In the Years 1740–1.* London, 1743.

Burchell, William J. *Travels in the Interior of Southern Africa.* 2 vols. London, 1822–1824.

Cannon, Walter F. 'The Impact of Uniformitarianism: Two Letters from John Herschel to Charles Lyell, 1836–1837.' *Proceedings of the American Philosophical Society.* vol. 105 (1961), pp. 301–314.

Carne, Joseph. 'On the relative age of the Veins of Cornwall.' *Transactions of the Royal Geological Society of Cornwall.* vol. 2 (1822), pp. 49–128.

Chapin, A. B. 'Junction of Trap and Sandstone, Wallingford, Conn.' *American Journal of Science and Arts.* vol. 27 (1835), pp. 104–112.

Chladni, Ernst Florenz Friedrich. 'Supplément au catalogue des météores, à la suite desquels des pierres ou des masses de fer sont tombées.' *Journal des mines.* vol. 26 (1809), pp. 79–80.

Conrad, Timothy Abbot. 'Observations on the Tertiary Strata of the Atlantic Coast.' *American Journal of Science and Arts.* vol. 28 (1835), pp. 104–111, 280–282.

Conybeare, William Daniel. 'Report on the Progress, Actual State, and Ulterior Prospects of Geological Science.' In *Report of the First and Second Meetings of the British Association for the Advancement of Science*, pp. 365–414. London, 1833.

Conybeare, William D., and Phillips, William. *Outlines of the Geology of England and Wales, with an Introductory Compendium of the General Principles of That Science, and Comparative Views of the Structure of Foreign Countries.... Part I.* London, 1822.

Cook, James. *A Voyage to the Pacific Ocean. Undertaken by the Command of His Majesty, for Making Discoveries in the Northern Hemisphere, to Determine the Position and Extent of the West Side of North America; Its Distance from Asia; and the Practicability of a Northern Passage to Europe. Performed under the Direction of Captains Cook, Clerke, and Gore, in His Majesty's Ships the Resolution and Discovery, in the Years 1776, 1777, 1778, 1779, and 1780. Vol. I and II Written by Captain James Cook... Vol. III by Captain James King... Published by Order of the Lords Commissioners of the Admiralty.* 3 vols. + atlas. London, 1784.

——. *A Voyage towards the South Pole, and round the World. Performed in His Majesty's Ships the Resolution and Adventure, in the Years 1772, 1773, 1774, and 1775. Written by James Cook, Commander of the Resolution. In Which Is Included, Captain Furneaux's Narrative of His Proceedings in the Adventure during the Separation of the Ships.* 2 vols. London, 1777.

[**Cooley, W. D.**] Review of *Coleccion de obras y documentos relativos a la historia antigua y moderna de las provincias del Rio de la Plata, ilustrados con notas y disertaciones* by Pedro de Angelis. *Edinburgh Review.* vol. 65 (1837), pp. 87–109.

Cortès, M., and Moreau de Jonnès, Alexandre. 'Mémoire sur la géologie des Antilles.' *Journal de physique, de chimie, d'histoire naturelle et des arts.* vol. 70 (1810), pp. 129–134.

Dampier, William. *A New Voyage round the World*... Vol. 1: *A New Voyage round the World. Describing particularly, the Isthmus of America, several Coasts and Islands in the West Indies, the Isles of Cape Verd, the Passage by Terra del Fuego, the South Sea Coasts of Chili, Peru, and Mexico; the Isle of Guam one of the Ladrones, Mindanao, and other Philippine and East-India Islands near Cambodia, China, Formosa, Luconia, Celebes, &c. New Holland, Sumatra, Nicobar Isles; the Cape of Good Hope and Santa Helena. Their Soil, Rivers, Labours, Plants, Fruits, Animals, and Inhabitants. Their Customs, Religion, Government, Trade, &c.* [1699] Vol. 2: *Voyages and Descriptions. In three Parts, viz. 1. A Supplement of the Voyage round the World, Describing the Countreys of Tonquin, Achin, Malacca, &c. their Product, Inhabitants, Manners, Trade, Policy, &c. 2. Two Voyages to Campeachy; with a Description of the Coasts, Product, Inhabitants, Logwood-Cutting, Trade, &c. of Jucatan, Campeachy, New-Spain, &c. 3. A Discourse of*

Trade-Winds, Breezes, Storms, Seasons of the Year, Tides and Currents of the Torrid Zone throughout the World: With an Account of Natal in Africk, its Product, Negro's &c. [1699] Vol. 3: *A Voyage to New Holland, &c. In the Year, 1699. Wherein are described, The Canary-Islands, the Isles of Mayo and St. Jago. The Bay of All Saints, with the Forts and Town of Bahia in Brasil. Cape Salvadore. The Winds on the Brasilian Coast. Abrohlo-Shoals. A Table of all the Variations observ'd in this Voyage. Occurrences near the Cape of Good Hope. The Course to New Holland. Shark's Bay. The Isles and Coast, &c. of New Holland. Their Inhabitants, Manners, Customs, Trade, &c., Their Harbours, Soil, Beasts, Birds, Fish, &c. Trees, Plants, Fruits, &c.* [1703]. 3 vols. 4th ed. London, 1699–1703.

Darby, William. *A Geographical Description of the State of Louisiana: Presenting a View of the Soil, Climate, Animal, Vegetable, and Mineral Productions; Illustrative of Its Natural Physiognomy, Its Geographical Configuration, and Relative Situation; With an Account of the Character and Manners of the Inhabitants. Being an Accompaniment to the Map of Louisiana.* Philadelphia, 1816.

Darwin, Charles. *Geological Observations on South America. Being the Third Part of the Geology of the Voyage of the Beagle, under the Command of Capt. Fitzroy, R.N. during the Years 1832 to 1836.* London, 1846.

——. *Geological Observations on the Volcanic Islands Visited during the Voyage of H.M.S. Beagle, together with Some Brief Notices of the Geology of Australia and the Cape of Good Hope. Being the Second Part of the Geology of the Voyage of the Beagle, under the Command of Capt. Fitzroy, R.N. during the Years 1832 to 1836.* London, 1844.

——. *Journal of Researches into the Geology and Natural History of the Various Countries Visited by H.M.S. Beagle, under the Command of Captain Fitzroy, R.N., from 1832 to 1836.* London, 1839. [Also published as Vol. 3 of Robert Fitzroy, ed., *Narrative of the Surveying Voyages of His Majesty's Ships Adventure and Beagle, between the Years 1826 and 1836 etc.*]

——. 'Observations on the Structure and Propagation of the genus Sagitta.' *Annals and Magazine of Natural History.* vol. 8 (1844), pp. 1–6 with plate.

——. Manuscripts. Cambridge University Library.

Darwin, Erasmus. *The Botanic Garden, A Poem. In Two Parts. Part I. Containing the Economy of Vegetation. Part II. The Loves of the Plants. With Philosophical Notes.* 2 vols. 4th ed. London, 1799.

Darwin, Francis, ed. *The Life and Letters of Charles Darwin, Including an Autobiographical Chapter.* 3 vols. 3rd ed. rev. London, 1888.

Daubeny, Charles. *A Description of Active and Extinct Volcanos; with Remarks on Their Origin, Their Chemical Phænomena, and the Character of Their Products, as*

Determined by the Condition of the Earth during the Period of Their Formation. London, 1826.

Davy, Humphry. 'On the corrosion of copper sheeting by sea water, and on methods of preventing this effect; and on their application to ships of war and other ships.' *Philosophical Transactions of the Royal Society of London,* vol. 114 (1824), pp. 151–158.

Debenham, Frank, ed. *The Voyage of Captain Bellingshausen to the Antarctic Seas 1819–1821.* 2 vols. London, 1945.

De La Beche, Henry Thomas. *A Geological Manual.* London, 1831.

Deluc, Jean André. *Geological Travels... Translated from the French Manuscript.* 3 vols. London, 1810–1811.

De Schauensee, Rodolphe Meyer. *The Species of Birds of South America and Their Distribution.* With the collaboration of Eugene Eisenmann. Narberth, Pennsylvania, 1966.

Ducatel, Julius T., and Alexander, John H. 'Report on a projected Geological and Topographical Survey of the State of Maryland.' *American Journal of Science and Arts.* vol. 27 (1835), pp. 1–38.

'Earthquake at Sea.' *Carmarthen Journal* (Wales), 3 April 1835.

'Earthquake at Sea.' *The Times* (London), 28 March 1835, p. 5.

Ehrenberg, Christian Gottfried. 'On the Origin of Organic Matter from simple Perceptible Matter, and on Organic Molecules and Atoms; together with some Remarks on the Power of Vision of the Human Eye.' In *Scientific Memoirs, Selected from the Transactions of Foreign Academies of Science and Learned Societies, and from Foreign Journals,* edited by Richard Taylor, vol. 1, pp. 555–583. London, 1837.

Falkner, Thomas. *A Description of Patagonia, and the Adjoining Parts of South America: Containing an Account of the Soil, Produce, Animals, Vales, Mountains, Rivers, Lakes, &c. of those Countries; the Religion, Government, Policy, Customs, Dress, Arms, and Language of the Indian Inhabitants; and Some Particulars relating to Falkland's Islands.* London, 1774.

Fitton, William. 'An Account of some Geological Specimens, collected by Captain P. P. King, in his Survey of the Coasts of Australia, and by Robert Brown, Esq., on the Shores of the Gulf of Carpentaria, during the Voyage of Captain Flinders.' In *Narrative of a Survey of the Intertropical and Western Coasts of Australia Performed between the Years 1818 and 1822 by Captain Phillip P. King... With an Appendix Containing Various Subjects Relating to Hydrography and Natural History,* by Phillip P. King, vol. 2, pp. 566–629. London, 1827.

Fitzroy, Robert, ed. *Narrative of the Surveying Voyages of His Majesty's Ships Adventure and Beagle, between the Years 1826 and 1836, Describing Their Examination of the Southern Shores of South America, and the Beagle's Circumnavigation of the Globe.* Vol. 1: *Proceedings of the First Expedition, 1826–1830, under the Command of Captain P. Parker King, R.N., F.R.S.* Vol. 2: *Proceedings of the Second Expedition, 1831–36, under the Command of Captain Robert Fitz-Roy, R.N.* [+ appendix]. Vol. 3: *Journal and Remarks. 1832–1836. by Charles Darwin, Esq., M.A.* 3 vols. + appendix. London, 1839.

Fox, Robert Were. 'On the electro-magnetic properties of metalliferous veins in the mines of Cornwall.' *Philosophical Transactions of the Royal Society of London.* vol. 120 (1830), pp. 399–414.

Gay, Claude. 'Aperçu sur les recherches d'histoire naturelle faites dans l'Amérique du sud, et principalement dans le Chili, pendant les années 1830 et 1831.' *Annales des sciences naturelles,* vol. 28 (1833), pp. 369–393.

Gebhard, John. 'On the Geology and Mineralogy of Schoharie, N.Y.' *American Journal of Science and Arts.* vol. 28 (1835), pp. 172–177.

Gould, John. 'Observations on the Raptorial Birds in Mr. Darwin's Collection, with characters of the New Species.' *Proceedings of the Zoological Society of London.* vol. 5 (1837), pp. 9–11.

——. 'On a New Rhea (*Rhea Darwinii*) from Mr. Darwin's Collection.' *Proceedings of the Zoological Society of London.* vol. 5 (1837), pp. 35–36.

——. *The Zoology of the Voyage of H.M.S. Beagle, under the Command of Captain Fitzroy, R.N., during the Years 1832–1836. Published with the Approval of the Lords Commissioners of Her Majesty's Treasury. Edited and Superintended by Charles Darwin...Naturalist to the Expedition. Part III. Birds: by John Gould...* 5 numbers. London, 1838–1841.

Great Britain. Hydrographic Office. *Index to Admiralty Published Charts,* Capt. F. J. Evans, Hydrographer. London, 1874.

Harris, Michael. *A Field Guide to the Birds of Galapagos.* London, 1974.

Helms, Anthony Zachariah. *Travels from Buenos Ayres, by Potosi, to Lima. With an Appendix, Containing Correct Descriptions of the Spanish Possessions in South America, Drawn from the Last and Best Authorities.* 2nd ed. London, 1807.

Henry, Samuel P. *Sailing Directions for Entering the Ports of Tahiti and Moorea.* London, 1852.

Henslow, John Stevens. 'Geological Description of Anglesea.' *Transactions of the Cambridge Philosophical Society.* vol. 1 (1821–1822), pp. 359–452.

Hoffman, Friedrich. *Geschichte der Geognosie, und Schilderung der Vulkanischen Erscheinungen.* Berlin, 1838.

Holland, Henry. 'Preliminary Dissertation on the History and Literature of Iceland.' In *Travels in the Island of Iceland, during the Summer of the Year MDCCCX [1810]*, by George Steuart Mackenzie, pp. 1–70. Edinburgh, 1811.

Humboldt, Alexander von. *Atlas géographique et physique des régions equinoxiales du nouveau continent.* Paris [F. Schoell], 1814.

——. *Fragmens de géologie et de climatologie asiatiques.* 2 vols. Paris, 1831.

——. *Personal Narrative of Travels to the Equinoctial Regions of the New Continent, during the Years 1799–1804, by Alexander de Humboldt, and Aimé Bonpland.* Translated by Helen Maria Williams. Vols. 1–2 [1822], 3rd ed. Vol. 3 [1822], 2nd ed. Vols. 4 [1819], 5 [1821], 6 [1826], 7 [1829], 1st ed. In 6 vols. London, 1819–1829.

——. *Political Essay on the Kingdom of New Spain. Containing Researches Relative to the Geography of Mexico, the Extent of Its Surface and Its Political Division into Intendancies, the Physical Aspect of the Country, the Population, the State of Agriculture and Manufacturing and Commercial Industry, the Canals Projected between the South Sea and Atlantic Ocean, the Crown Revenues, the Quantity of the Precious Metals Which Have Flowed from Mexico into Europe and Asia, Since the Discovery of the New Continent, and the Military Defence of New Spain.* Translated by John Black. 4 vols. London, 1811.

Hunter, John. *An Historical Journal of the Transactions at Port Jackson and Norfolk Island, with the Discoveries which have been made in New South Wales and in the Southern Ocean, since the publication of Phillip's Voyage, compiled from the Official Papers; Including the Journals of Governors Phillip and King, and of Lieut. Ball; and the Voyages From the first Sailing of the Sirius in 1787, to the Return of that Ship's Company to England in 1792. By John Hunter, Esqr. Post Captain in His Majesty's Navy.* London, 1793.

Hutton, James. *Theory of the Earth, with Proofs and Illustrations. In Four Parts.* 2 vols. Edinburgh, 1795. [Vol. 3. Edited from manuscript by Sir Archibald Geikie. London, 1899.]

Isabelle, Arsène. *Voyage à Buénos-Ayres et à Porto-Alègre, par la Banda-Oriental, les Missions d'Uruguay et la province de Rio-Grande-do-Sul. (De 1830 à 1834.) Suivi de considérations sur l'état du commerce français à l'extérieur, et principalement au Brésil et au Rio-de-la-Plata....* Havre, 1835.

Johnson, Alfredo William. *The Birds of Chile and Adjacent Regions of Argentina, Bolivia and Peru.* 2 vols. Buenos Aires, 1965–1967.

Juan, George, and Ulloa, Antonio de. *A Voyage to South America: Describing at Large the Spanish Cities, Towns, Provinces, &c. on That Extensive Continent.*

Undertaken, by Command of the King of Spain, by Don George Juan, and Don Antonio de Ulloa, Captains of the Spanish Navy. Translated...with Notes and Observations; and an Account of the Brazils by John Adams, Esq. 2 vols. 4th ed. London, 1806.

Kendal, Edward. 'Account of the Island of Deception, one of the New Shetland Isles. Extracted from the private Journal of Lieutenant Kendal, R.N., embarked on board his Majesty's sloop Chanticleer, Captain Forster, on a scientific voyage....' *Journal of the Royal Geographical Society.* vol. 1 (1832), pp. 62–66.

Kotzebue, Otto von. *A Voyage of Discovery, into the South Sea and Beering's Straits, for the Purpose of Exploring a North-East Passage, Undertaken in the Years 1815–1818, at the Expense of His Highness the Chancellor of the Empire, Count Romanzoff, in the Ship Rurick, under the Command of the Lieutenant in the Russian Imperial Navy, Otto von Kotzebue.* Translated by H. E. Lloyd. 3 vols. London, 1821.

Labillardière, Jacques Julien Houton de. *Relation du voyage à la recherche de La Pérouse, fait par ordre de l'Assemblée constituante, pendant les années 1791, 1792, et pendant la 1ère. et la 2de. année de la République françoise.* 2 vols. Paris, an VIII [1800].

La Condamine, Charles Marie de. *A Succinct Abridgment of a Voyage Made within the Inland Parts of South-America; from the Coasts of the South-Sea, to the Coasts of Brazil and Guiana, down the River of Amazons: As it was read in the Public Assembly of the Academy of Sciences at Paris, April 28, 1745.* London, 1747.

La Pérouse, Jean François Galaup de. *A Voyage round the World Performed in the Years 1785, 1786, 1787, and 1788, by the Boussole and Astrolabe, under the Command of J. F. G. de la Pérouse: Published by Order of the National Assembly, under the Superintendence of L. A. Milet-Mureau.* 2 vols. + atlas. London, 1798–1799.

Lesson, René Primevère, and Garnot, Prosper. *Voyage autour du monde, exécuté par ordre du Roi, sur la corvette de Sa Majesté, La Coquille, pendant les années 1822, 1823, 1824 et 1825, sous le Ministère et conformément aux instructions de S.E.M. le marquis Clermont-Tonnerre, ministre de la marine; et publié sous les auspices de son excellence Mgr le Cte de Chabrol, ministre de la marine et des colonies, par M. L. I. Duperrey,...Zoologie, par MM. Lesson et Garnot.* 2 vols. + atlas. Paris, 1826–1830.

Lister, Joseph Jackson. 'Some Observations on the Structure and Functions of tubular and cellular Polypi, and of Ascidiæ.' *Philosophical Transactions of the Royal Society of London.* vol. 126 (1834), pp. 365–388.

Lyell, Charles. *Principles of Geology, Being an Attempt to Explain the Former Changes of the Earth's Surface, by Reference to Causes Now in Operation.* 3 vols. London, 1830–1833.

——. *Principles of Geology, Being an Inquiry How Far the Former Changes of the Earth's Surface Are Referable to Causes Now in Operation.* 4 vols. 4th ed. London, 1835; 4 vols. 5th ed. London, 1837.

Mackenzie, George Steuart. *Travels in the Island of Iceland, during the Summer of the Year MDCCCX [1810]*. Edinburgh, 1811.

Malaspina, Alessandro. 'Tablas de latitudes y longitudes de los principales puntos del Rio de la Plata, nuevamente arregladas al meridiano que pasa por lo mas occidental de la Isla de Ferro.' In *Coleccion de obras y documentos relativos a la historia antigua y moderna de las provincias del Rio de la Plata* by Pedro de Angelis, vol. 6. Buenos Aires, 1837.

——. *Viaje político-científico alrededor del mundo por las corbetas Descubierta y Atrevida al mando de los capitanes de navío D. Alejandro Malaspina y Don José de Bustamante y Guerra desde 1789 á 1794, publicado con una introducción por Don Pedro de Novo y Colson....* Madrid, 1885.

Marsden, William. *The History of Sumatra, Containing an Account of the Government, Laws, Customs, and Manners of the Native Inhabitants, with a Description of the Natural Productions, and a Relation of the Ancient Political State of That Island.* 3rd ed. London, 1811.

Mathison, Gilbert Farquhar. *Narrative of a Visit to Brazil, Chile, Peru, and the Sandwich Islands, during the Years 1821 and 1822. With Miscellaneous Remarks on the Past and Present State, and Political Prospects of Those Countries.* London, 1825.

Mawe, John. *Travels in the Gold and Diamond Districts of Brazil; Describing the Methods of Working the Mines, the Natural Productions, Agriculture, and Commerce, and the Customs and Manners of the Inhabitants: To Which Is Added a Brief Account of the Process of Amalgamation Practised in Peru and Chili.* A new ed. London, 1825.

Michell, John. 'Conjectures concerning the Cause, and Observations on the Phænomena of Earthquakes; particularly of that Great Earthquake of the First of November 1755, which proved so fatal to the City of Lisbon, and whose Effects were felt as far as Africa, and more or less throughout almost all Europe.' *Philosophical Transactions of the Royal Society of London.* vol. 51 (1760), pp. 566–634.

Miers, John. *Travels in Chile and La Plata, including Accounts respecting the Geography, Geology, Statistics, Government, Finances, Agriculture, Manners and Customs, and the Mining Operations in Chile. Collected during a Residence of Several Years in These Countries.* 2 vols. London, 1826.

Mitscherlich, Eilhert. 'On Artificial Crystals of Oxide of Iron.' *Edinburgh Journal of Natural and Geographical Science.* vol. 2 (1830), p. 302.

Molina, Juan Ignacio. *Compendio de la historia geografica, natural y civil del reyno de Chile, escrito en italiano por el abate don Juan Ignacio Molina*. . . Part 1: *Compendio . . . Primera parte, que abraza la historia geografica y natural, traducida en español por don Domingo Joseph de Arquellada Mendoza*. . . Part 2: *Compendio de la historia civil del reyno de Chile. . . Parte segunda, traducida al español y aumentada con varias notas por don Nicolas de la Cruz y Bahamonde.* 2 vols. Madrid, 1788–1795.

Morris, John, and Sharpe, Daniel. 'Description of Eight Species of Brachiopodous Shells from the Palaeozoic Rocks of the Falkland Islands.' *Quarterly Journal of the Society of London.* vol. 2 (1846), pp. 274–278.

Morton, Samuel George. 'Notice of the fossil teeth of Fishes of the United States, the discovery of the Galt in Alabama, and a proposed division of the American Cretaceous Group.' *American Journal of Science and Arts.* vol. 28 (1835), pp. 276–278.

Mulhall, Michael G. *The English in South America.* Buenos Aires and London, 1878.

Murchison, Roderick Impey. *The Silurian System, Founded on Geological Researches in the Counties of Salop, Hereford, Radnor, Montgomery, Caermarthen, Brecon, Pembroke, Monmouth, Gloucester, Worcester, and Stafford; with Descriptions of the Coal-Fields and Overlying Formations.* London, 1839.

'Notice of the Transactions of the Geological Society of Pennsylvania, Part I.' *American Journal of Science and Arts.* vol. 27 (1835), pp. 347–355.

Orbigny, Alcide Dessalines d'. *Voyage dans l'Amérique méridionale (le Brésil, la république orientale de l'Uruguay, la République argentine, la Patagonie, la république du Chili, la république de Bolivia, la république du Pérou), exécuté pendant les années 1826, 1827, 1828, 1829, 1830, 1831, 1832 et 1833, par Alcide d'Orbigny. . . Ouvrage dédié au Roi, et publié sous les auspices de M. le ministre de l'instruction publique*. . . . 9 vols. Paris, 1835–1847.

Owen, Richard. *The Zoology of the Voyage of H.M.S. Beagle, under the Command of Captain Fitzroy, R.N., during the Years 1832 to 1836. Published with the Approval of the Lords Commissioners of Her Majesty's Treasury. Edited and Superintended by Charles Darwin. . . Naturalist to the Expedition. Part I. Fossil Mammalia: by Richard Owen*. . . . 4 numbers. London, 1838–1840.

Owen, William F. W. *Narrative of Voyages to Explore the Shores of Africa, Arabia, and Madagascar; Performed in H.M. Ships Leven and Barracouta, under the Direction of Captain W. F. W. Owen, R.N. by Command of the Lords Commissioners of the Admiralty.* 2 vols. London, 1833.

Pallas, Peter Simon. *Travels through the Southern Provinces of the Russian Empire, in the Years 1793 and 1794. Translated from the German of P. S. Pallas....* 2 vols. London, 1802–1803.

Parish, Woodbine. *Buenos Ayres, and the Provinces of the Rio de la Plata: Their Present State, Trade, and Debt. With Some Account from Original Documents of the Progress of Geographical Discovery in Those Parts of South America during the Last Sixty Years.* London, 1839.

——. 'Notice as to the supposed Identity of the large Mass of Meteoric Iron now in the British Museum, with the celebrated Otumpa Iron described by Rubin de Celis in the Philosophical Transactions for 1786.' *Philosophical Transactions of the Royal Society of London.* vol. 124 (1834), pp. 53–54.

Pernety, Antoine Joseph. *Journal historique d'un voyage fait aux îles Malouïnes en 1763 & 1764, pour les reconnoître, & y former un établissement; et de deux voyages au détroit de Magellan, avec une rélation sur les Patagons.* 2 vols. Berlin, 1769.

Péron, François. *Voyage de découvertes aux terres australes. Historique.* Vol. 1: *Voyage de découvertes aux terres australes, exécuté par ordre de Sa Majesté l'empereur et roi, sur les corvettes le Géographe, le Naturaliste, et la goëlette le Casaurina, pendant les années 1800, 1801, 1802, 1803 et 1804; publié par décret impérial, sous le ministère de M. de Champagny, et rédigé par M. F. Péron, naturaliste de l'expédition....* Vol. 2: *Voyage de découvertes aux terres australes, exécuté sur les corvettes le Géographe, le Naturaliste, et la goëlette le Casuarina, pendant les années 1800, 1801, 1802, 1803 et 1804; publié par ordre de Son Excellence le ministre secrétaire d'état de intérieur.... Rédigé en partie par feu F. Péron, et continué par M. Louis Freycinet....* 2 vols. + atlas. Paris, 1807–1816.

Phillip, Arthur. *The Voyage of Governor Phillip to Botany Bay; with an Account of the Establishment of the Colonies of Port Jackson & Norfolk Island; compiled from Authentic Papers, which have been obtained from the several Departments. To which are added the Journals of Lieuts. Shortland, Watts, Ball, & Capt. Marshall with an Account of their New Discoveries....* London, 1789.

Phillips, William. *An Elementary Introduction to the Knowledge of Mineralogy: Comprising Some Account of the Characters and Elements of Minerals; Explanations of Terms in Common Use; Descriptions of Minerals, with Accounts of the Places and Circumstances in Which They Are Found; and Especially the Localities of British Minerals.* 3rd ed. London, 1823.

Playfair, John. *Illustrations of the Huttonian Theory of the Earth.* Edinburgh, 1802.

[**'Proteus'**]. 'The Bahama Islands.' *United Service Journal and Naval and Military Magazine.* vol. 3 (1834), pp. 215–226.

Quoy, Jean René Constantin, and Gaimard, Paul. 'Observations zoologiques faites à bord de l'Astrolabe, en Mai 1826, dans le Détroit de Gibraltar.' *Annales des sciences naturelles.* vol. 10 (1827), pp. 5–22, 172–193, 225–239.

Rackett, Thomas. 'Observations on *Cancer salinus.*' *Transactions of the Linnean Society of London.* vol. 11 (1815), pp. 205–206.

Rogers, Henry D. 'On the Falls of Niagara and the reasonings of some authors respecting them.' *American Journal of Science and the Arts.* vol. 27 (1835), pp. 326–335.

Ross, John. *A Voyage of Discovery, Made under the Orders of the Admiralty, in His Majesty's Ships Isabella and Alexander, for the Purpose of Exploring Baffin's Bay, and Inquiring into the Probability of a North-West Passage.* London, 1819.

Roussin, Albin-Reine. *Le Pilote du Brésil, ou Description des côtes de l'Amérique méridionale, situées entre l'île Santa-Catharina et celle de Maranhaõ; cartes et plans de ces côtes et instructions pour naviguer dans les mers du Brésil, composé sur les documents recueillis dans la compagne hydrographique...exécutée en 1819 et 1820 sur la corvette la Bayadère et le brig le Favori...par M. le Bon Roussin,...*Paris, 1826.

Scoresby, Jr., William. *An Account of the Arctic Regions, with a History and Description of the Northern Whale-fishery.* 2 vols. Edinburgh, 1820.

Scrope, George Poulett. *Considerations on Volcanos, the Probable Causes of Their Phenomena, the Laws Which Determine Their March, the Disposition of Their Products, and Their Connexion with the Present State and Past History of the Globe; Leading to the Establishment of a New Theory of the Earth.* London, 1825.

Shepard, Charles U. 'On the Strontianite of Schoharie, (N.Y.) with a Notice of the Limestone Cavern in the same place.' *American Journal of Science and Arts.* vol. 27 (1835), pp. 363–370.

Shuttleworth, Nina L. Kay. *A Life of Sir Woodbine Parish, K.C.H., F.R.S., 1796–1882....*London, 1910.

Smith, Andrew. *Illustrations of the Zoology of South Africa; Consisting Chiefly of Figures and Descriptions of the Objects of Natural History Collected during an Expedition into the Interior of South Africa, in the Years 1834, 1835, and 1836; Fitted Out by "The Cape of Good Hope Association for Exploring Central Africa:" together with a Summary of African Zoology, and an Inquiry into the Geographical Ranges of Species in that Quarter of the Globe.* 5 parts. London, 1838–1849.

Sturt, Charles. *Two Expeditions into the Interior of Southern Australia during the Years 1828, 1829, 1830, and 1831: With Observations on the Soil, Climate, and*

General Resources of the Colony of New South Wales. By Capt. Charles Sturt, 39th Regt.... 2 vols. London, 1833.

Temple, Edmond. *Travels in Various Parts of Peru, Including a Year's Residence in Potosi*. 2 vols. London, 1830.

Totten, Joseph G. 'Descriptions of some Shells, belonging to the Coast of New England.' *American Journal of Science and Arts*. vol. 28 (1835), pp. 347–353.

Ulloa, Antonio de. *Noticias americanas: entretenimientos físico-históricos sobre la América Meridional, y la Septentrional oriental: comparacion general de los territorios, climas y producciones en las tres especies vegetal, animal y mineral; con una relacion particular de los Indios de aquellos paises, sus costumbres y usos, de las petrificaciones de cuerpos marinos, y de las antigüedades. Con un discurso sobre el idioma, y conjeturas sobre el modo con que pasáron los primeros pobladores*. 2nd ed. Madrid, 1792.

Volney, Constantine François. *Voyage en Syrie et Égypte pendant les années 1783, 1784 et 1785*. 2 vols. 2nd ed. rev. Paris, 1787.

Webster, John W. *A Description of the Island of St. Michael, Comprising an Account of Its Geological Structure; with Remarks on the Other Azores or Western Islands. Originally Communicated to the Linnæan Society of New-England*. Boston, 1821.

Webster, William H. B. *Narrative of a Voyage to the Southern Atlantic Ocean, in the Years 1828, 29, 30, Performed in H.M. Sloop Chanticleer, under the Command of the Late Captain Henry Foster, F.R.S. &c., by Order of the Lords Commissioners of the Admiralty*. 2 vols. London, 1834.

Williams, John. *A Narrative of Missionary Enterprises in the South Sea Islands; with Remarks upon the Natural History of the Islands, Origin, Languages, Traditions, and Usages of the Inhabitants*. London, 1837.

Wilson, Leonard G. *Charles Lyell: The Years to 1841. The Revolution in Geology*. New Haven and London, 1972.

Index to Persons Named in the Red Notebook

Page references are to the original notebook

Subject Index to the Red Notebook

Page references are to the original notebook

Geographical Index to the Red Notebook

Page references are to the original notebook

Library of Congress Cataloging in Publication Data

DARWIN, CHARLES ROBERT, 1809–1882
The red notebook of Charles Darwin.

Includes indexes.
1. Darwin, Charles Robert, 1809–1882. 2. Natural history. 3. Naturalists—England—Biography.
I. Herbert, Sandra. II. Title.
QH31.D2A37 575′.0092′4 78-74215
ISBN 0-8014-1226-9